Coatings to Improve Optoelectronic Devices

Coatings to Improve Optoelectronic Devices

Selected Articles Published by MDPI

MDPI • Basel • Beijing • Wuhan • Barcelona • Belgrade • Manchester • Tokyo • Cluj • Tianjin

This is a reprint of articles published online by the open access publisher MDPI (available at: http://www.mdpi.com). The responsibility for the book's title and preface lies with Alicia de Andrés, who compiled this selection.

For citation purposes, cite each article independently as indicated on the article page online and as indicated below:

LastName, A.A.; LastName, B.B.; LastName, C.C. Article Title. *Journal Name* **Year**, *Article Number*, Page Range.

ISBN 978-3-03928-334-7 (Pbk)
ISBN 978-3-03928-335-4 (PDF)

© 2020 by the authors. Articles in this book are Open Access and distributed under the Creative Commons Attribution (CC BY) license, which allows users to download, copy and build upon published articles, as long as the author and publisher are properly credited, which ensures maximum dissemination and a wider impact of our publications.

Contents

Preface to "Coatings to Improve Optoelectronic Devices" vii

Nengduo Zhang, Jian Sun and Hao Gong
Transparent p-Type Semiconductors: Copper-Based Oxides and Oxychalcogenides
Reprinted from: *Coatings* 2019, 9, 137, doi:10.3390/coatings9020137 1

Nuria Novas, Alfredo Alcayde, Dalia El Khaled and Francisco Manzano-Agugliaro
Coatings in Photovoltaic Solar Energy Worldwide Research
Reprinted from: *Coatings* 2019, 9, 797, doi:10.3390/coatings9120797 29

Ashraf Uddin, Mushfika Baishakhi Upama, Haimang Yi and Leiping Duan
Encapsulation of Organic and Perovskite Solar Cells: A Review
Reprinted from: *Coatings* 2019, 9, 65, doi:10.3390/coatings9020065 51

Kensuke Nishioka, So Pyay Moe and Yasuyuki Ota
Long-Term Reliability Evaluation of Silica-Based Coating with Antireflection Effect for Photovoltaic Modules
Reprinted from: *Coatings* 2019, 9, 49, doi:10.3390/coatings9010049 69

Grégory Barbillon
Fabrication and SERS Performances of Metal/Si and Metal/ZnO Nanosensors: A Review
Reprinted from: *Coatings* 2019, 9, 86, doi:10.3390/coatings9020086 77

Manuela Proença, Marco S. Rodrigues, Joel Borges and Filipe Vaz
Gas Sensing with Nanoplasmonic Thin Films Composed of Nanoparticles (Au, Ag) Dispersed in a CuO Matrix
Reprinted from: *Coatings* 2019, 9, 337, doi:10.3390/coatings9050337 91

Yuan-Chang Liang and Che-Wei Chang
Preparation of Orthorhombic WO_3 Thin Films and Their Crystal Quality-Dependent Dye Photodegradation Ability
Reprinted from: *Coatings* 2019, 9, 90, doi:10.3390/coatings9020090 103

Doga Bilican, Samer Kurdi, Yi Zhu, Pau Solsona, Eva Pellicer, Zoe H. Barber, Alan Lindsay Greer, Jordi Sort and Jordina Fornell
Epitaxial Versus Polycrystalline Shape Memory Cu-Al-Ni Thin Films
Reprinted from: *Coatings* 2019, 9, 308, doi:10.3390/coatings9050308 115

Yuki Kameya and Hiroki Yabe
Optical and Superhydrophilic Characteristics of TiO_2 Coating with Subwavelength Surface Structure Consisting of Spherical Nanoparticle Aggregates
Reprinted from: *Coatings* 2019, 9, 547, doi:10.3390/coatings9090547 125

Olaf Stenzel, Steffen Wilbrandt, Sven Stempfhuber, Dieter Gäbler and Sabrina-Jasmin Wolleb
Spectrophotometric Characterization of Thin Copper and Gold Films Prepared by Electron Beam Evaporation: Thickness Dependence of the Drude Damping Parameter
Reprinted from: *Coatings* 2019, 9, 181, doi:10.3390/coatings9030181 135

Simon Bublitz and Christian Mühlig
Absolute Absorption Measurements in Optical Coatings by Laser Induced Deflection
Reprinted from: *Coatings* 2019, 9, 473, doi:10.3390/coatings9080473 149

Preface to "Coatings to Improve Optoelectronic Devices"

This selection of reviews and articles is focused on coatings and thin films with applications in optoelectronics, such as photovoltaics, photocatalysis, and light-based sensors and phenomena. The investigations pursue the optimal composition, crystalline structure, and morphology able to deliver in term of the different functionalities sought.

Most of the optoelectronic devices, in particular, the massively growing transparent display market, require transparent conducting materials (TCMs) as electrodes. There are different well-established n-type conducting oxides (ITO, AZO, FTO, ...) with high optical transparency, where electrons are the charge carriers. However, the development of p-type transparent electrodes—TCMs with hole carriers—is very challenging and, nonetheless, would be extremely relevant for OLED or perovskite solar cell devices. One review summarizes the novelties of Cu-based oxide films with delafossite structure, as well as oxychalcogenides films to be used as transparent p-type electrodes. The electric conduction mechanisms and the correlations with structure and doping are discussed [1].

According to the study included in this compendium [2], the main topics in solar coatings—as deduced from an analysis of the publications in the field—are the search for materials for photovoltaics, especially inorganic and organic thin films, as well as ways to enhance light trapping and reduce reflection losses. These problems are tackled by different approaches, such as layer nanostructuration, incorporation of nanoparticles, or coating with photonic crystals. Very recently the efficiency, cost-effectiveness, and simplicity of fabricating of metal–organic hybrid perovskite layers for photovoltaic devices have revolutionized the field. However, very relevant issues limit their actual applications: their poor stability under ambient conditions and their short photodegradation periods. A review of the degradation pathways as well as the possibilities offered by different types of coatings to encapsulate the devices [3] is of paramount relevance to the field. On the other hand, the efficiency of well-established technologies, such as those based on silicon or on $Cu(In,Ga)Se_2$ (CIGS), can be increased by appropriate silica coating, acting both as an antireflection and antisoiling layer [4].

The enhancement of Raman signal for the detection of analytes is of paramount importance because the technique is nondestructive and specific, since the vibration modes are characteristic of each molecule or crystal. Investigation on the capabilities of plasmonic resonance of metallic nanoparticles for SERS is still a hot topic. Here, the combination of metallic nanoparticles and nanostructured semiconductors is reviewed for chemical and biological molecule sensing. Different nanostructures for the semiconductor, Si or ZnO, such as nanowires, nanorods, hollow nanospheres, nanocones, or nanoneedles, are used to deposit silver or gold nanoparticles and their enhancement factors are collected and discussed [5]. Another interesting approach is the fabrication of nanoplasmonic films consisting of silver, gold, or bi-metallic Ag–Au nanoparticles dispersed in CuO films for the detection of gas molecules, O_2, through the modification of the surface plasmon resonance peak position in the transmittance optical spectra [6].

A central problem in thin films is the correlation between their structure and morphology and the final performance of the system, discussed for TCMs in [1]. An example is the relevance of crystallinity and the concentration of oxygen vacancies in the photoactivated properties of WO_3 films. Both aspects are crucial for the electronic structure and band gap that finally determine the light-induced degradation ability of organic dyes [7]. Also in this direction, high crystallinity seems to be required to obtain shape memory behavior in Cu-Al-Ni thin films grown on MgO (001) substrates: the martensitic transformation is only obtained in epitaxial films [8]. On its side, the control of morphology of a coated layer brings the possibility to modify the device properties. Aggregated TiO_2 nanoparticles with submicron sizes provide a graded refractive index from air to the TiO_2 layer that enhances the optical transmittance and confers a superhydrophilic character maintaining the photocatalytic activity [9].

To finalize this compendium, two reports on very relevant aspects on the fundamental characterization of thin films are included. Optical properties are directly related to the refractive index of the media, however, the optical constants for thin films, in many cases, are not reported and bulk data are not always applicable. The dependence of Drude damping as a function of the film thickness is reported for gold and copper films, and the models discussed [10]. The laser-induced deflection technique is introduced to provide absolute absorption data that allow obtaining individual absorption and scattering data and, thus, the real and imaginary parts of the refractive index of coatings [11].

Alicia de Andrés

Review

Transparent p-Type Semiconductors: Copper-Based Oxides and Oxychalcogenides

Nengduo Zhang [1,2,†], Jian Sun [3,†] and Hao Gong [1,*]

1. Department of Materials Science and Engineering, National University of Singapore, Singapore 117576, Singapore; e0011485@u.nus.edu
2. NUS Graduate School for Integrative Sciences and Engineering, National University of Singapore, Singapore 117456, Singapore
3. Faculty of Materials Science and Chemistry, China University of Geoscience, Wuhan 430074, China; sunjian@cug.edu.cn
* Correspondence: msegongh@nus.edu.sg
† These authors contributed equally to this work.

Received: 31 January 2019; Accepted: 16 February 2019; Published: 20 February 2019

Abstract: While p-type transparent conducting materials (TCMs) are crucial for many optoelectronic applications, their performance is still not satisfactory. This has impeded the development of many devices such as photovoltaics, sensors, and transparent electronics. Among the various p-type TCMs proposed so far, Cu-based oxides and oxychalcogenides have demonstrated promising results in terms of their optical and electrical properties. Hence, they are the focus of this current review. Their basic material properties, including their crystal structures, conduction mechanisms, and electronic structures will be covered, as well as their device applications. Also, the development of performance enhancement strategies including doping/co-doping, annealing, and other innovative ways to improve conductivity will be discussed in detail.

Keywords: p-type semiconductors; transparent oxides; delafossite; oxychalcogenide

1. Introduction

Transparent conducting oxides (TCOs) possess high electrical conductivity and good optical transparency in the visible light range, and they have been intensively studied over the past few decades due to their important roles in electronic industries including photovoltaic cells (PV cells), touch screen displays, solid-state sensors, organic light-emitting diodes (OLEDs), and liquid crystal displays [1–3]. Currently, in the electronics industry, many different materials, especially impurity-doped materials, are used as TCOs for the applications. Three major materials are In_2O_3 doped with Sn (ITO), ZnO doped with Al (AZO), Fluor tin oxide or SnO_2–F (FTO), and SnO_2 doped with Sb (ATO) [4,5], Among them, ITO, with its great electrical conductivity around 1000 S·cm^{-1} and optical transparency greater than 80%, has a market share higher than 97% [6,7].

However, the aforementioned materials all belong to n-type TCOs, which show n-type conductivity. The popularity of n-type TCOs arises from their preferable electronic properties [8]. For n-type TCOs, the electrons as charge carriers move in the conduction band minimum (CBM). The CBM is largely formed by the spatially spread metal-s orbitals, resulting in a well-dispersed CBM and high electron mobility. In addition, the low formation energy of native intrinsic defects induces a high electron concentration and stable n-type electrical conductivity after impurity doping [9]. On the contrary, the development of high-performance p-type TCOs is very challenging [10,11], and it remains a grand challenge for researchers to solve. Indeed, it is very difficult to find p-type TCO with high conductivity due to the inherent low mobility of holes in the oxides. As can be seen in Figure 1, the effective mass of a hole is relatively higher than that of an electron [12]. This is caused by the valence band maximum (VBM) for hole transport being mainly formed by highly localized oxygen 2p orbitals, leading to low hole mobility. Moreover, the low formation energy of intrinsic donor defect and relatively high acceptor formation energy limit the number of hole carriers [11].

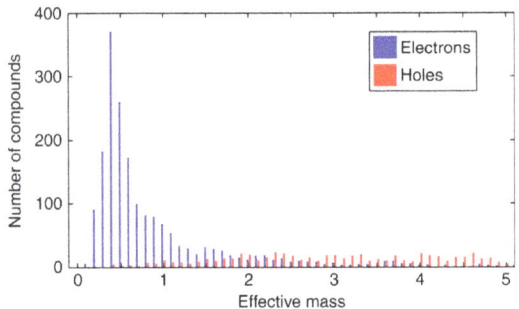

Figure 1. Effective mass distribution of both electrons and holes in selected binary and ternary oxides. Reproduced with permission from Ref. [12]. Copyright 2013 Springer Nature.

On the other hand, p-type TCOs are extremely important in many applications. For example, a p-type TCO, with its larger work function than its n-type counterpart, is more suitable for OLED devices [13,14]. However, with performance sacrifice, many of the current OLED applications use ITO as the transparent electrode instead of a p-type anode material due to the lack of high-performance p-type TCO [13,14]. In the perovskite solar cells, a hole transport layer (HTL) usually is made of the popular semi-metallic organic material poly(3,4-ethylenedioxythiophene):poly(styrenesulfonate) (PEDOT:PSS), and this has been shown to negatively affect the stability of the device, which is a serious issue for perovskite solar cells [15–18]. Moreover, in the field of transparent electronics, wide gap oxides are no longer passive components in the device. Instead, transparent oxides play an essential role as active layers, similar to Si in current semiconductor industry. After achieving high-performance p-type TCOs, transparent p–n junctions and a complementary metal oxide semiconductor (CMOS) could be fabricated for various applications with many more advantages than its unipolar transistors such as lower heat generation, higher circuit density, and lower energy consumption [19–21]. It is also believed that transparent electronics will become one of the most promising technologies for next-generation flat panel display [22], and the forecast shows that the transparent display would help create a $87.2 billion market by 2025, as can be seen in Figure 2 [2].

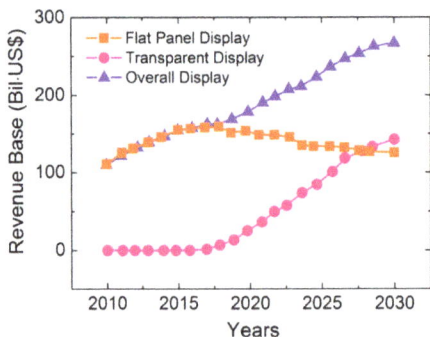

Figure 2. Transparent display market. Reproduced with permission from Ref. [2]. Copyright 2012 Wiley.

In order for p-type TCOs to achieve reasonable electrical properties, in 1997, Kawazoe et al. introduced a concept called "Chemical Modulation of the Valence Band" (CMVB), as a way to deal with the aforementioned problem [23,24]. The delafossite structured CuAlO$_2$ was prepared and discussed in that work. In delafossites, Cu$^+$ has a closed shell of 3d^{10} orbitals which would hybridize with O 2p orbitals and modify the VBM and make the hole more delocalized so that it increases the hole mobility [25]. This hybridization is possible because the energy level of Cu 3d^{10} is close to that of O 2p. Subsequently, several other types of delafossites with the structure CuMO$_2$ (M = Al [26–28], Cr [29–31], Ga [32–34], In [35–37], Sc [38–40], Y [41–43], Fe [44–46], and B [47–49]) and delafossite-related oxide SrCu$_2$O$_2$ [50–52] have been investigated. In 2001, the delafossite with a rather high p-type conductivity of 220 S·cm^{-1} was achieved by Mg-doped CuCrO$_2$, but the transparency is not satisfying [30]. Many of these delafossites have been used in the applications such as OLEDs as a hole injection layer (HIL) or hole transport layer (HTL) due to their suitable work function, good stability, and preferable electronic properties [53,54]. While a lot of attention has been paid to such p-type TCOs, it seems very challenging for delafossites to attain the high conductivity and high transparency that would make them comparable to current n-type TCOs [55]. The rather low conductivity is caused by their deep acceptor level [56], and polaronic nature [43,57]. In some cases, the transparency of a delafossite is limited by the interband transitions [58]. The band-engineering CMVB technique was later used on other Cu$^+$-based materials, especially materials with chalcogens such Cu$_2$S, Cu$_2$Se, and Cu$_2$Te [59]. Even though chalcogen-based TCOs have greater hole mobility due to their stronger hybridization with Cu 3d states by the chalcogens than oxygen [10], they suffer greatly from their narrow band gap, which limits their visible light transparency [60]. Therefore, it appears that a high-performance TCO should include oxygen for a large band gap and chalcogen for a high hole mobility. Subsequently, layered oxychalcogenides have been developed to demonstrate a large band gap and high hole mobility due to the hybridization of Cu and chalcogens. Layered oxychalcogenide has alternate layers of [LnO]$^+$ (Ln = La, Y, Pr, etc.) and [CuCh]$^-$ (Ch = S, Se, Te, etc.), in which the [CuCh]$^-$ is essential for its p-type conductivity and band gap [61,62]. It has been reported that the conduction band minimum (CBM) and the valence band maximum (VBM) comprise a Cu–Ch hybridized band and Cu–Ch anti-bonding band, respectively [13,63]. This forms a direct band gap that allows a direct transition of charge carriers, which is more useful in applications than indirect band-gap materials [10,62,63]. While oxychalcogenides have promising properties that enable them to be used as future transparent p-type materials, their hole mobility is around 10^0 cm$^2\cdot$V$^{-1}\cdot$s^{-1} or less. For applications such as thin-film transistors (TFTs), which require materials with high mobility, cuprous oxide (Cu$_2$O), with its high hole mobility greater than 2.0×10^2 cm$^2\cdot$V$^{-1}\cdot$s^{-1}, seems more suitable [64], even though it has a small band gap (2.0–2.5 eV) that is only transparent for part of the visible light range [65–67]. The high mobility of Cu$_2$O originates from the rather delocalized VBM, which is formed mainly by Cu d states, and the VB is formed by the hybridization of O 2p and Cu 3d orbitals [68,69]. In this review, we will focus on the discussion of three types of materials, including delafossites, oxychalcogenides, and copper oxides.

2. Preparation and Properties of Materials

2.1. Delafossite (CuMO$_2$)

Before the realization of CuAlO$_2$ delafossite (band gap > 3 eV) as a transparent p-type oxide [24], p-type conductivity in oxides with large band gaps was too low to be used in any industrial applications. As briefly mentioned above in the introduction, the low electrical conductivity is the result of the general electronic structure in the metal oxides. This is because oxygen 2p orbitals are generally at a much lower energy level than that of the metallic atoms in ionic metal oxides, leading to a highly localized and deep valence band that is formed by oxygen 2p orbitals. Holes generated in the oxides would be attracted and localized by oxygen ions, leading to very low hole mobility [8,23]. To tackle the problem of low hole mobility, the highly localized VB or VBM need to be modified to be more dispersed. To achieve such band modification, a design strategy named "Chemical Modulation of the Valence Band" (CMVB) was proposed in 1997 [23,24]. The cation selected by this strategy should possess the electronic configuration such that its highest shell electrons should be situated at similar energy levels as that of oxygen 2p orbitals. As shown in Figure 3a, this would introduce chemical bonds and make the newly formed VB or VBM more spatially dispersed. In addition, the selected cation should also have a closed shell valence state to prevent coloration in the transparent oxides caused by d–d transitions. After carefully selecting the right cations, it was found that both Cu^+ and Ag^+ possess suitable electronic configuration of $d^{10}s^0$, and they are at a similar energy level as that of oxygen 2p orbitals [23]. When Cu^+ or Ag^+ cations are brought to oxygen ions, chemical bonds with a certain degree of covalency would be formed, and the VBM would be formed by the anti-bonding levels of the resulting band formation (as can be observed in Figure 3a). Besides the overlap of the orbitals to create dispersed VBs or VBMs, the crystal structure is also important for the transparent oxide performance for two reasons. Firstly, the tetrahedral coordination of oxygen would ensure all eight of its electrons were involved in the four sigma bonds, as can be seen in Figure 3b [70]. As the non-bonded electrons that are normally present in oxides would result in the generated holes having low mobility, the absence of non-bonded electrons of oxygen would lead to a higher hole mobility. Secondly, the crystal structure of the delafossite is also advantageous to its optical properties. For the prototypical delafossite $CuAlO_2$, it has a layered structure with two alternating arrays comprising of Cu and the edge-sharing distorted MO_6 octahedral unit along the vertical axis, and the two layers are connected by the O–Cu–O bonds [43]. This reduces the cross-linking between Cu^+ as compared to that in Cu_2O, which would decrease the interaction between the d^{10} electrons of Cu^+ ions. Since it has been reported that such d^{10} electrons interaction between neighboring atoms would decrease the band gap such as the case in Cu_2O where interaction among three-dimensional $3d^{10}$ electrons is present, the layered delafossite would improve the optical transparency [11,24]. Therefore, the delafossite based on Cu^+ has been predicted to be a potential high performance candidate for p-type TCOs in terms of its electrical and optical properties. In addition, the doping of divalent atoms at the site of trivalent metal M is believed to greatly enhance the p-type conductivity of delafossites [11], even though the effect of size of the trivalent metal atom on the conductivity trend still remains under intense debate [30,71]. It should be noted that even though both Cu^+ and Ag^+ have a suitable electronic configuration of $d^{10}s^0$, Cu^+-based delafossites are usually preferred due to the following two reasons [49]. Firstly, Ag-based delafossites are difficult to be synthesized partially due to the low free energy of forming Ag_2O, and decomposition will occur at 300 °C into O and Ag [72]. Secondly, it was shown experimentally that Ag-based delafossites exhibit lower conductivity compared to Cu-based delafossites due to unfavorable mixing between the Ag 4d and O 2p states [71,73–75].

Delafossites with the formula $CuMO_2$ have two polytypes, namely a rhombohedral 3R and hexagonal 2H type, depending on the stacking sequence of the Cu and double layer MO_2. For the rhombohedral 3R type, the alternative Cu layer has the same orientation between each MO_2 layer within three layers, and each unit cell has three layers with the space group $R\bar{3}m$ (No. 166) (Figure 3b). For the hexagonal 2H type, each unit cell contains two layers with MO_2 and Cu layers with 180^0 offset for each alternating Cu layer with the space group $P6_3/mmc$ (No. 194) (Figure 3c). It has been reported that certain delafossites such as $CuAlO_2$, $CuGaO_2$, and $CuInO_2$ generally prefer the rhombohedral 3R type, while other delafossites such as $CuScO_2$, $CuYO_2$, and $CuLaO_2$ normally adopt a hexagonal 2H form [76]. In addition, the former type of delafossite has an indirect band gap, and the many latter ones possess a direct band gap [43,77,78]. It should be noted that the relative energy difference between the two polytypes is generally rather small [79,80], so the relative phase stability for each delafossite is still under debate [81]. Different synthesis routes and preparation conditions might result in different symmetries, and it is not uncommon to find the prepared delafossite with a mixture of both 3R and 2H polymorphs [82–84]. So far, most of the delafossites that have been used for p-type TCO applications are reported to possess rhombohedral 3R structures [49,81,85–87].

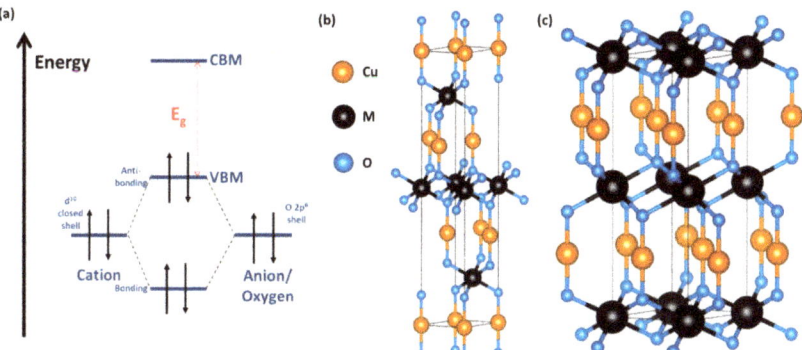

Figure 3. (**a**) Schematic diagram of the interaction between cation with a d^{10} closed shell and oxygen anion, the anti-bonding formed by the hybridization give the valence band maximum (VBM) of the delafossites. Crystal structure of the delafossite of (**b**) rhombohedral 3R type and (**c**) hexagonal 2H type.

As mentioned above, $CuAlO_2$ is the first material in the delafossite family to be prepared as a p-type transparent oxide back in 1997 [24], even though $CuAlO_2$ was already illustrated to possess p-type conductivity in 1984 [88]. Since its first introduction as a promising p-type TCO, a lot of research effort has been put on the development of delafossites with good electrical and optical properties. Since 1997, many thin film fabrication methods have been used to prepare transparent p-type $CuMO_2$ films, including vacuum-based routes such as radio-frequency magnetron sputtering (RFMS) [89,90], pulsed laser deposition (PLD) [32,91], chemical vapor deposition (CVD) [26,29], atomic layer deposition (ALD) [92,93] and non-vacuum based routes such as the hydrothermal/solvothermal method [94,95], sol-gel process [27,96], the spin-coating method [97,98], and the spray-coating method [99,100]. Each thin film fabrication method has its own advantages. For example, RFMS is relatively cost-efficient among all the vacuum-based methods, and is compatible with industrial applications for large scalability [101,102]; PLD is better at conserving the stoichiometry ratio in the target and producing high-quality epitaxy thin films [103–105]; and ALD has good control of both elemental composition and thin film morphology [106,107]. While many high-performance delafossites have been prepared via these state-of-art vacuum-based approaches, the chemical solution-based method has attracted more and more research attention recently, as it allows for low-cost fabrication, roll-to-roll capability, low-temperature synthesis, high throughput, and the possibility for special nanostructures [108–110]. Today, the above-mentioned thin film approaches are used in various research groups to prepare $CuMO_2$, as discussed briefly below. The electrical and optical properties of the selected high-performance delafossites are tabulated in Table 1.

$CuAlO_2$ was firstly prepared as a transparent p-type oxide in 1997 by using PLD at a substrate temperature of 700 °C [24]. The deposited thin film of 500 nm belongs to the rhombohedral 3R type, and it has a p-type conductivity of around 1 S·cm^{-1} and an optical band gap of 3.5 eV with a poor transmittance around 30% [24]. After this, the same group investigated the electronic properties of $CuAlO_2$ in more detail, and they obtained thin film with conductivity around 0.3 S·cm^{-1} and transparency of 70% via the PLD approach [111]. It should be noted that the delafossite $CuMO_2$ phase is not trivial to achieve due to the unstable oxidation state of Cu^+ and the complex phase diagram of Cu, M, and O [112]. Later studies has been carried out to achieve phase-pure $CuAlO_2$, in order to eliminate common impurities such as $CuAl_2O_4$ and CuO [91]. In 2000, Gong et al. developed a Cu–Al–O-based semiconductor thin film via the CVD method that is industrial compatible, and the p-type conductivity is as high as 2 S·cm^{-1} [26]. As an attempt to further improve the electrical conductivity of intrinsic $CuAlO_2$, high-quality nanocrystalline oxide particles were prepared via the hydrothermal method, and the deposited thin film possesses a conductivity of 2.4 S·cm^{-1} with a transparency at the 550-nm wavelength around 55% [94]. Similar to other semiconducting oxides, doping is usually an effective way to increase electrical conductivity. Various theoretical studies has been carried out on delafossite $CuAlO_2$ to study its doping possibility, and Group II (alkaline earth) elements were suggested as suitable dopants [113,114]. Detailed experimental evidence was shown by Dong et al., where 2% Mg improves the $CuAlO_2$ by three orders, although the final conductivity is only 0.083 S·cm^{-1} [115]. Besides the hard work to improve the optoelectronic properties, the conducting mechanism and material physics were examined by both experiments and theoretical studies [87,116]. It has been shown that the close-packed Cu$^+$ alternative layers act as the hole transport path [117]. Due to its anisotropic structural properties, the ab-planes have a hole mobility that is 25 times lower than the c-axis, where the mobility along the c-axis is around 3.0 cm$^2 \cdot V^{-1} \cdot s^{-1}$ [118]. The p-type conductivity of $CuAlO_2$ was proven to be originated from the Cu vacancies and Cu on Al antisites in the oxides, which have a rather small formation energy, and Cu vacancy is a shallow-level defect with an ionization energy of 0.7 eV [87,89,118]. In contrast, O interstitials were found to be not responsible for the p-type conductivity, as they are deep-level defects and relatively unstable [87,119]. In addition, it was generally believed that the conduction mechanism for $CuAlO_2$ and other delafossites is via an activated small polaronic hopping mechanism, in which distorted lattice and trapped holes interact and move between Cu sites, leading to relatively low hole mobility [43,57,87,116]. In order to achieve a higher hole mobility of p-type conductivity, Yao et al. recently used an innovative way to enhance the hole mobility of $CuAlO_2$ with an ultra-high Hall mobility of 39.5 cm$^2 \cdot V^{-1} \cdot s^{-1}$ [120]. Polycrystalline $CuAlO_2$ deposited on the Si substrate via RFMS was incorporated with different amount of non-isovalent CuO. The atomic ratio of Cu^{2+}/Cu$^+$ is varied, from 4.2% and 9.3%, to 22.8%, and the optical band gaps were changed from 3.79 eV and 3.72 eV to 3.46 eV, respectively. The high mobility achieved is found to be caused by the hybridization of Cu–O dimers in the $CuAlO_2$ lattice. The Cu–O dimers have a VBM of 0.6 eV, which is higher than that of $CuAlO_2$, so it would raise the VBM of the alloyed $CuAlO_2$/CuO. Moreover, the more diverse VBM of CuO, which was formed mainly by Cu^{2+} 3d^9 bands, would modulate the VBM of $CuAlO_2$/CuO and make it more delocalized, leading to a high hole mobility. Indeed, the top gate TFT made by this alloyed $CuAlO_2$/CuO achieved good device performance with a high field effect mobility of 0.97 cm$^2 \cdot V^{-1} \cdot s^{-1}$, and the schematic drawing of the TFT device with its detailed device performance is shown in Figure 4 [120].

Table 1. Summary of selected delafossites with their material properties including thickness (*d*), conductivity (σ), transmittance (*T*), and optical band gap (E_g). ALD: atomic layer deposition, CVD: chemical vapor deposition; PLD: pulsed laser deposition; RFMS: radio-frequency magnetron sputtering.

Delafossites	Fabrication	*d* (nm)	σ (S·cm^{-1})	*T* (%)	E_g (eV)	Ref.
$CuAlO_2$	PLD	230	0.3	70	3.50	[111]
$CuAlO_2$	Hydrothermal	420	2.4	55	3.75	[94]
$CuAlO_2$	RFMS	–	2.7×10^{-2}	–	3.79	[120]
$CuScO_2$	RFMS	110	30	40	3.30	[38]
$CuCrO_2$:Mg	RFMS	250	220	30	3.10	[30]
$CuCrO_2$:Mg	ALD	120	217	70	3.00	[93]
$CuCrO_2$:Mg	Spray coating	155	1	80	3.08	[121]
$CuCrO_2$:Zn	Sol-gel	205	0.47	55	3.05	[122]
$CuCrO_2$:N	RFMS	290	17	55	3.19	[123]
$CuCrO_2$:Mg/N	RFMS	150	278	69	3.52	[124]
$CuCrO_2$	CVD	140	17	50	3.20	[125]
$Cu_{0.66}Cr_{1.33}O_2$	CVD	200	102	35	3.22	[126]
$CuCrO_2$	Spray coating	90	12	55	–	[127]
$CuBO_2$	PLD	200	1.65	85	4.50	[47]
$CuGaO_2$	PLD	500	6.3×10^{-2}	80	3.60	[32]
$CuInO_2$:Ca	PLD	170	2.8×10^{-3}	70	3.90	[35]
$SrCu_2O_2$:K	PLD	120	4.8×10^{-2}	70	3.30	[50]
$SrCu_2O_2$	PLD	220	4.5×10^{-2}	80	3.30	[128]

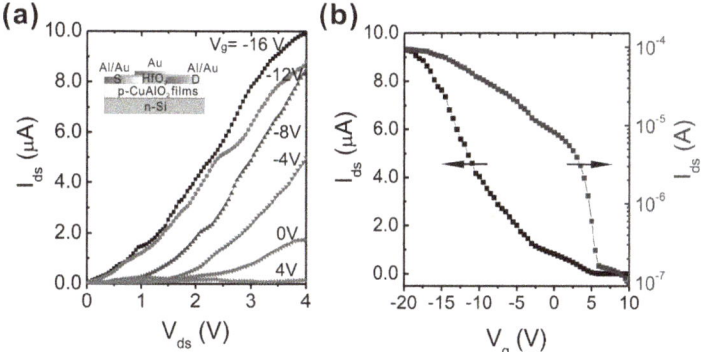

Figure 4. (a) The change of I_{ds} against V_{ds} at different V_g values; (b) The change of I_{ds} against V_g at V_{ds} at 3 V. Inset in (a) is the schematic drawing of the thin-film transistor (TFT). Reproduced with permission from Ref. [120]. Copyright 2012 AIP Publishing.

Compared to the prototypical delafossite $CuAlO_2$, $CuCrO_2$ has been found to possess a higher p-type conductivity, especially after divalent atom doping such as Mg. Moreover, it differs from other delafossites by having better hybridization between M 3d states and O 2p states, and possessing better dupability [129–131]. However, it has a poorer optical transparency due to the d-d transitions in Cr [58]. On the other hand, the intrinsic or stoichiometric $CuCrO_2$ has rather low p-type conductivity of 10^{-4} S·cm^{-1}, while the divalent atom doped or Cu-deficient $CuCrO_2$ possesses a high hole conductivity [126]. Until 2018, the $CuCrO_2$ doped with Mg that was prepared by Nagarajan et al. in 2001 held the p-type conductivity record for delafossites of 220 S·cm^{-1}, but it also had a poor optical transparency of 30% [30]. An attempt to improve the transparency for such a film decreased the conductivity to around 1 S·cm^{-1}. After the report of the high conductivity by that work, no following publications could be found for this system with a high conductivity. Many other dopants, including Fe, Ca, Ni, Sr, Ba, and Zn have since been incorporated into the $CuCrO_2$, but their p-type conductivity is much lower than 220 S·cm^{-1} [122,132–135]. Besides the substitution at the trivalent M sites, the N atom has been used as a substitute for the O atom in $CuCrO_2$. It was found that the p-type conductivity increased by more than three orders to 17 S·cm^{-1} compared to the intrinsic state [123]. By combining both cation and anion substitution, a new record for p-type conductivity in delafossites has been reported to be 278 S·cm^{-1} in 2018 [124]. In this study, Mg and N dopants were introduced into the $CuCrO_2$ thin film, which served to substitute Cr and O, respectively. With the doping concentration of 2.5% Mg and a 40% $N_2/(N_2+Ar)$ ratio during sputtering, the $CuCrO_2$ film had the record-high conductivity of 278 S·cm^{-1} and optical transparency of 69%. A theoretical study using first-principle calculation on Mg and N co-doping has also supported the effectiveness of such an approach to improve p-type conductivity [136]. Possibly because this paper was published just recently, no other reports on such a high conductivity for this system can be found. Besides the Mg–N doping, other co-doping strategies have also been studied, including Mg–Fe [133], Mg–S [137], and Ag–Mg [138], but the electrical and optical properties are not comparable to the study on Mg–N doping [124].

Recently, Cu-deficient $CuCrO_2$ has been reported to possess a relatively high p-type conductivity, and their optical and electrical properties are tabulated in Table 1 [109,125–127,139]. In 2017, Popa et al. prepared highly Cu-deficient $Cu_{0.66}Cr_{1.33}O_2$ with a conductivity higher than 100 S·cm^{-1} without any intentional extrinsic doping, and this appears to be the highest conductivity value out of all the delafossites without extrinsic doping [126]. However, no other papers for a similarly high conductivity for this system have been reported since then. The origin of this high conductivity was proposed to be the high Cu deficiency, and the missing of part of a Cu plane was observed for the as-deposited $Cu_{0.66}Cr_{1.33}O_2$ film. After high-temperature annealing at 900 °C, the $CuCrO_2$ film restored its stoichiometric ratio with Cr_2O_3 precipitates. The electrical conductivity dropped significantly as the Cu defect was healed, which suggests that the Cu deficiency defect is metastable [126]. Besides the quest for only high-performance CuCrO$_2$, more cost-efficient polymer substrates that are compatible with low-temperature chemical solution methods have been developed recently, and all of the CuCrO$_2$ films that were fabricated depicted a Cu deficiency with a high p-type conductivity [109,127,139]. It is known that high-temperature processing for these p-type oxides makes them unsuitable for flexible applications using polymer substrates [140]. In 2018, Wang et al. investigated a low-process temperature at 180 °C to prepare CuCrO$_2$ delafossites, and p-n diode and organic photovoltaic devices have been prepared successfully [139]. Also, several solution-based CuCrO$_2$ thin films have been fabricated and investigated for their applications, including p-n junctions, thermoelectric devices, and organic solar cells [141–143]. Recently, Nie developed a solution-based CuCrO$_2$ thin-film transistor (TFT) device [109]. A high on/off ratio of around 10^5 and good field effect mobility of 0.59 cm$^2 \cdot$V$^{-1} \cdot$s^{-1} was achieved.

The good performance of CuCrO$_2$ encouraged research groups to study the effect of trivalent M in CuMO$_2$ delafossites in detail. It has been explained before that the presence of Cu–O–Cu linkage in Cu$_2$O gives a high hole mobility in Cu$_2$O because of the hybridization of Cu and O states, but the trivalent M is involved in the Cu–O–M–O–Cu linkage in delafossites. This would lead to a lower mobility in delafossites. On the other hand, the higher conductivity of CuCrO$_2$ might be caused by the hybridization of Cr 3d states in the linkage [42]. It has also been theoretically shown that Cr 3d states would hybridize with O 2p states to form more dispersed VBM, leading to lower hole effective mass [131]. However, different conducting mechanisms of Mg-doped CuCrO$_2$ have been proposed over the years, including the p-type conduction through Cu^{2+}/Cu$^+$ [134] or Cr^{4+}/Cr^{3+} [144]. Unfortunately, the exact role of trivalent M on the conductivity is still not fully understood yet. This complexity is partially caused by the difficulty to locate the exact Cr 3d state energy position [134,145]. In 2013, Yokobori et al. performed resonant photoelectron spectroscopy (PES) on CuCrO$_2$:Mg, and the Cu–Cr 3d–3p resonant PES spectra showed that VBM was primarily composed of Cr 3d states [146]. This work suggested that the holes generated in the oxides would also be delocalized to Cr sites, leading to a larger hole mobility. However, this report also showed that holes would be localized onto Cu sites according to X-ray absorption spectroscopy (XAS). Nevertheless, it is generally believed that the VBM is mainly formed by Cu 3d states, and this is consistent with the X-ray photoelectron spectroscopy (XPS) studies [129]. Another theoretical study by Scanlon et al. suggested that the high p-type conductivity of CuCrO$_2$ originates from the well hybridization of Cr 3d states with O 2p states, which would improve hole conductivity [56]. Indeed, more experimental and theoretical studies should be carried out to investigate the mechanism of conductivity enhancement for CuCrO$_2$ and compare it to other delafossites. Another controversial issue for CuCrO$_2$ is whether it is a direct or indirect band-gap oxide [29,147,148]. It has been often reported to have an indirect band gap around 2.5–2.8 eV and a direct band gap around 3.0–3.3 eV [29], although it was also shown to possess only a fundamental band gap of 3.2 eV [147]. A later detailed simulation study has proven that CuCrO$_2$ indeed possesses an indirect band gap [56].

By building on the studies of trivalent M effects, CuBO$_2$ was later proposed as a potential p-type TCO [47]. This is because the B in CuBO$_2$ has the smallest ionic size in the delafossite family, which will significantly increase the optical band gap according to the theoretical studies [77,149]. Snure et al. prepared CuBO$_2$ film by using PLD for the first time, and a decent p-type conductivity of 1.65 S·cm^{-1} with a high optical transparency higher than 85% at the 550-nm wavelength was achieved [47]. However, it was noted that an indirect band gap at 2.2 eV was observed, and transmittance dropped greatly below the 550-nm wavelength. Later studies have been carried out to prepare nanocrystalline or microstructure CuBO$_2$ via chemical solution methods, although the performance in terms of the electrical conductivity and optical transmittance is not comparable to the CuCrO$_2$ films, as discussed before [150–152]. Another popular delafossite under intense research investigation is CuGaO$_2$, which was firstly prepared by Yanagi et al. in 2000 [33]. Early studies showed that CuGaO$_2$ has rather good optical transparency (higher than 80%) with decent electrical conductivity [32]. Due to its good optical properties and ease of nanostructure synthesis, CuGaO$_2$ has been studied for the applications of p-type dye-sensitized or perovskite solar cells [54,153–156]. For example, CuGaO$_2$ has been used as a HTL in the perovskite solar cell CH$_3$NH$_3$PbI$_{3-x}$Cl$_x$ with a high power conversion efficiency (PCE) of 18.51% due to the good transparency, relatively high mobility, and proper low-lying VBM [54]. The structure of both the material CuGaO$_2$ and the whole solar cell device with its band diagram is shown in Figure 5. As can be seen from the schematic energy band diagram, holes in the organic semiconducting perovskite layers can be transported to the CuGaO$_2$ layer effectively due to their well-matched energy levels. Another promising delafossite is CuInO$_2$, as it had been found to be a bipolar semiconductor. By either Ca or Sn doping, CuInO$_2$ could be made into a p-type or n-type TCO, respectively, and all of the oxide transparent p-n homojunctions could be fabricated [35–37,149]. The rather unique bipolar property could be explained by the theory of equilibrium doping and the doping limit rule [149,157–160]. The VBM and CBM positions directly influence the extent of self-compensation. A higher VBM could allow for easier p-type doping, while a lower CBM could also induce easier n-type doping. In the case of CuInO$_2$, it has a high VBM similar to other delafossites, contributing to its p-type dopability. However, the CBM lies at a much lower energy level compared to other delafossites (1.48 eV lower than CuAlO$_2$) due to the low-lying In states, leading to its n-type dupability [149]. It is also interesting to note that delafossites with larger ionic sizes of trivalent metal M such as CuLaO$_2$, CuYO$_2$, and CuScO$_2$ could accommodate additional oxygen atoms [161]. Unlike other delafossites with smaller M sizes such as CuCrO$_2$, oxygen interstitials start to contribute to the p-type conductivity in those delafossites with larger M ionic sizes [38,162]. Treating thin films such as CuScO$_2$ at a high temperature (400–500 °C) in oxygen-rich atmosphere could incorporate a large amount of oxygen (CuScO$_{2.5}$) and enhance the conductivity [38].

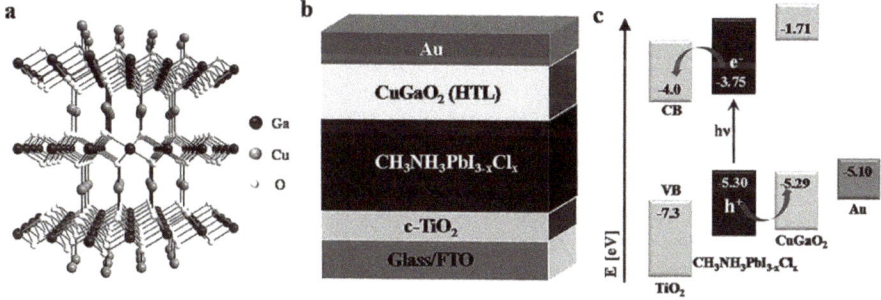

Figure 5. (a) Crystal structure of CuGaO$_2$, (b) schematic illustration of the solar cell device, and (c) band diagram of the materials in the device. Reproduced with permission from Ref. [54]. Copyright 2017 Wiley.

Apart from the CuMO$_2$ delafossites discussed so far, SrCu$_2$O$_2$ is another important Cu-based one that is used as a p-type transparent semiconductor. It was proposed on the same basis as the selection rules of delafossites, and it has similar working principles for the VBM formation with delafossites or Cu$_2$O [50]. It has a tetragonal crystal structure with O–Cu–O dumbbell bonding (as shown in Figure 6a), which is similar to delafossites or Cu$_2$O. These dumbbell bonds form the one-dimension chain in a zigzag manner, so the transparency could be improved [50]. From both the first-principle calculations and experiments, SrCu$_2$O$_2$ was found to possess a direct band gap of 3.3 eV, and the VBM is formed by Cu 3d–4sp orbitals and O 2p orbitals, leading to higher hole mobility [163]. Dopant K was found to be useful for the partial substitution at the Sr site to generate holes, and the conductivity improved from 3.9×10^{-3} to 4.8×10^{-2} S·cm^{-1}, while the transparency at a visible range remained largely unchanged [50]. Due to its relatively low processing or deposition temperature, which are as low as 300 °C, an all-oxide transparent p-n junction (shown in Figure 6b) was made by ZnO/SrCu$_2$O$_2$ with a device transparency higher than 70% and an ideality factor of 1.62 [164], and several UV/near-UV emitting diodes were constructed [52,165,166]. Even though the direct band gap SrCu$_2$O$_2$ has a relatively strong photoluminescence (PL) peak at around 500 nm, the light emitted by heteroepitaxial grown ZnO/SrCu$_2$O$_2$ was shown to be originated from the ZnO layer with an emission peak around 380 nm [166].

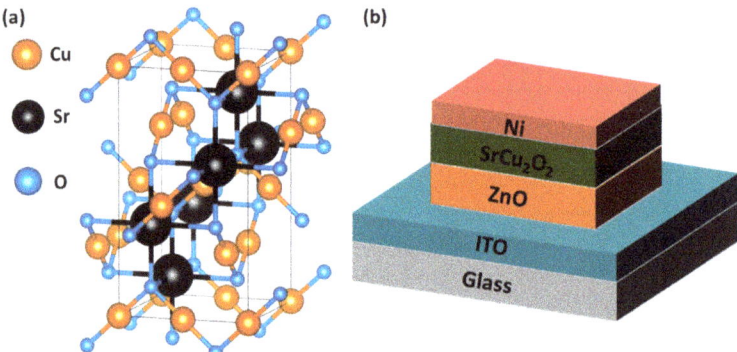

Figure 6. (**a**) Crystal structure of SrCu$_2$O$_2$; (**b**) Schematic drawing of a typical p-n junction by ZnO/SrCu$_2$O$_2$.

2.2. Oxychalcogenides

As briefly discussed before, the delafossite has the limitation of relatively low mobility, and its conduction mechanism is based on polaronic hopping. It had been proposed that oxychalcogenide, with its prototypical form, LaCuOS, would achieve better electrical performance with higher mobility [87]. As the VBM is mainly formed by the well hybridization between Cu 3d orbitals and S 3p orbital states, this could achieve higher mobility compared to delafossites as more spatially dispersed VBM is formed [167]. Since the introduction of the first oxychalcogenide (LaCuOS) as a p-type transparent material in 2000 by Ueda et al., many forms of oxychalcogenides with the formula LnCuOCh (Ln = La, Y, Pr, etc., and Ch = S, Se, Te, etc.) have been investigated [168]. It was later reported that doping LaCuOS with Sr could increase its conductivity to 0.26 S·cm^{-1} [169]. The band gap would decrease when La was replaced with other Ln elements (Ln = Pr and Nd) [170]. Similarly, the band gap would decrease when sulfide (S) was replaced by other chalcogens such as Se and Te, and the p-type conductivity would increase at the same time [171]. Among the various dopants that have been used to date, Mg dopants delivered rather good results. It has been reported that LaCuOSe doped with Mg exhibited a p-type conductivity of 910 S·cm^{-1} without an indication of transparency [63]. Another interesting oxychalcogenide is [Cu$_2$S$_2$][Sr$_3$Sc$_2$O$_5$], where [Sr$_3$Sc$_2$O$_5$] replaced the [La$_2$O$_2$] group. This undoped oxychalcogenide with a wide band gap of 3.14eV has a p-type conductivity of 2.8 S·cm^{-1}, which is even higher than the undoped CuBO$_2$ [47] that has a high intrinsic p-type conductivity in the delafossite family [172]. The source of this high undoped conductivity is its extremely high hole mobility of 150 cm^2·V^{-1}·s^{-1}, which is higher than any other p-type TCOs, and even higher than the highest mobility of n-type TCOs [173]. Strangely, no follow-up experimental works have been reported on this material since that work.

LnCuOCh belongs to the tetragonal system with the space group P4/nmm, and it has a layered structure with alternating [Ln$_2$O$_2$]$^{2+}$ and [Cu$_2$S$_2$]$^{2-}$ layers along the c-axis, as shown in Figure 7a. In each layer, the structure of [Ln$_2$O$_2$]$^{2+}$ has a PbO type, while that of the [Cu$_2$S$_2$]$^{2-}$ layer has an anti-PbO type [172,174]. It has also been reported that the [Ln$_2$O$_2$]$^{2+}$ layer is much more resistive to hole carriers than the [Cu$_2$S$_2$]$^{2-}$ layer, and the [Cu$_2$S$_2$]$^{2-}$ layer is indeed the hole transporting path [62]. Moreover, the two-dimensional characteristics demonstrated by the layered LnCuOCh could suggest its similarity to the artificial multiple quantum wells (MQWs). From the energy-band calculations, it should be regarded as a natural MQW due to its hole confinement effect between the insulating $[Ln_2O_2]^{2+}$ layers, which leads to many novel electronics applications [62]. This two-dimensional quantum confinement effect leading to the relatively large exciton binding energy of around 50 meV for LnCuOCh, resulting in good photoluminescence (PL) performance at room temperature, which also indicates its direct band-gap characteristic [170,175].

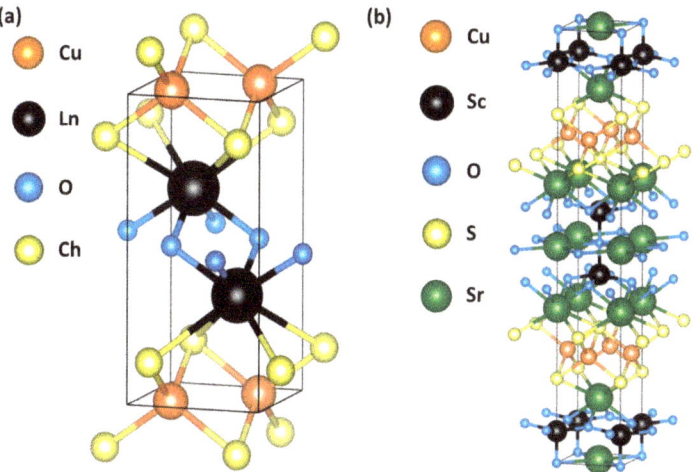

Figure 7. Crystal structure of (**a**) LnCuOCh (Ln = La, Y, Pr, etc., and Ch = S, Se, Te, etc.) and (**b**) $[Cu_2S_2][Sr_3Sc_2O_5]$.

One of the advantages of oxychalcogenides compared to other oxides is the tunability of their physical properties [14]. By changing the chalcogen elements from S through Se to Te, the conductivity, band gap, and other physical properties can be tuned. Usually, oxides are more robust compared to chalcogenides, resulting in a limited ability to be tuned. Moreover, as the quaternary layered structure, oxychalcogenides have more degrees of freedom for exploring more possibilities. In the series of LaCuOCh, the band gaps decrease from LaCuOS of 3.14 eV via LaCuOSe of 2.8 eV to LaCuOTe of 2.4 eV [176]. This is caused by the energy-level differences among the Ch p orbitals, as the VBM is formed by the hybridization between the Cu $3d$ and Ch p orbitals (as shown in Figure 8) [177]. Therefore, LaCuOTe has the highest degree of hybridization for its VBM, leading to the highest hole mobility among all LaCuOCh. However, it should be noted that LaCuOTe has an indirect band gap and no room temperature PL signal, and both LaCuOS and LaCuOSe have direct band gaps with intense PL signals [171]. On the other hand, for LnCuOS series, the band gaps decrease from LaCuOS of 3.14 eV via PrCuOS of 3.03 eV to NdCuOS of 2.98 eV [174]. Unlike the case for Ch substitution, as discussed before, the band gap does not correlate with the energy levels of Ln orbitals in the $[Ln_2O_2]^{2+}$ layer directly [170]. Instead, the substitution of Ln from La via Pr to Nd decreases the a-axis parameter, and this would reduce the Cu–Cu distance. The reduction in the distance would lead to higher Cu $4s$ interaction, resulting in the Cu $4s$ related band broadening, and thus a lower band gap that is similar to the case of YCuOSe (shown in Figure 8) [178].

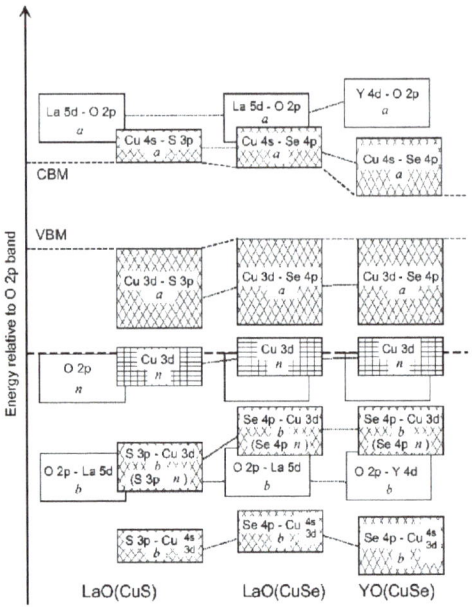

Figure 8. Schematic diagram of the band structures of LaCuOS, LaCuOSe, and YOCuSe from the first-principle calculation. Reproduced with permission from Ref. [178]. Copyright 2007 AIP Publishing.

In 2000, Ueda et al. firstly introduced LaCuOS as a potential p-type oxychalcogenide in the form of polycrystalline thin film [168]. The intrinsic LaCuOS after post-annealing is a partially transparent semiconducting film with a conductivity of 1.2×10^{-2} S·cm^{-1} and transmittance higher than 70%. After 5% Sr doping, the conductivity was improved more than one order to 2.6×10^{-1} S·cm^{-1}. After optimizing the Sr doping concentration to 3%, the conductivity was further increased to 20 S·cm^{-1} [179]. Similarly, the incorporation of a bivalent Sr atom was also found to be effective for polycrystalline LaCuOSe and other LnCuOCh materials for conductivity enhancement [174,180]. However, Sr doping was shown to be ineffective for epitaxial LaCuOCh film, and Mg was found to be able to improve the conductivity in this case [181]. As mentioned before, the Mg-doped expitaxial film LaCuOSe by reactive solid-phase epitaxy (R-SPE) is still holding the p-type conductivity record of 910 S·cm^{-1} among all the p-type TCOs [63]. The less than 1% Mg doping in the 40-nm thin film attained degenerate conduction, which cannot be achieved by Ca or Sr doping [168,179]. Due to the high hole concentration of higher than 10^{21} cm^{-3}, free carrier absorption (FCA) was present in such film, and optical absorption in the visible light and infrared region occurs [182]. From the Drude model, the effective mass of the holes could be calculated from FCA [183]. Moreover, such a high hole concentration is only present in such thin films with thicknesses of 40 nm, and is absent in thicker films of 150 nm, and hole concentration is much higher than the Mg doping concentration [63,175]. As the Cu concentration varies according to depth in thinner films (40-nm thickness), this suggested that the origin of hole generation is related to Cu vacancies [63]. Another interesting and surprising feature of this high conductivity film is that the mobility after doping only dropped a little, and this is because the hole carriers are transferred to the [Cu$_2$Se$_2$] layer, so the holes are not scattered by the charged impurities [184]. However, it suffers from poor transparency due to the lower band gap of 2.8 eV and subgap optical absorption ranging from 2.0 to 2.8 eV [13]. On the other hand, Mg-doped LaCuOSe, which is a p-type degenerate semiconductor, does not experience a Moss–Burstein (MB) shift, which is a blue shift of the optical absorption [185]. Instead, subgap absorption is observed, and its origin is not yet well understood [13]. This has been suggested by the observation that the acceptor defects in LaCuOSe keep the Fermi level above the VBM, leading to the absence of an MB shift effect [173]. Moreover, from the theoretical calculation and perspective of cation radius, Sr should be the optimal dopant, followed by Ca for LaCuOCh, instead of Mg [173]. While the explanation for this phenomenon is not clear yet, it could be related to the origin of the conduction mechanism in LnCuOCh. It has been shown that the source of p-type conductivity in LnCuOCh is Cu deficiency, where Cu vacancies are the acceptor defects [186]. The presence of Mg does not directly contribute a hole by substituting Ln sites; instead, it serves to enhance the Cu vacancies, leading to higher hole concentrations [13]. The electrical and optical properties of selected oxychalcogenides are tabulated in Table 2. Besides the improvement of the properties for LnCuOCh materials, several devices were also fabricated to demonstrate their good optoelectronic performances [187,188]. Hiramatsu et al. prepared a LED in the blue–ultraviolet region by using LaCuOSe as the p-type material and InGaZn$_5$O$_8$ as the n-type material [187]. This p-n heterojunction has demonstrated a strong and sharp electroluminescence at the 430-nm wavelength at room temperature. Later, Yanagi et al. used LaCuOSe doped with Mg as the hole injection layer/electrode for OLEDs [188]. The LaCuOSe:Mg that was used in that work offered advantages over the widely used ITO, including a lower hole injection barrier, higher current drivability, and lower threshold voltage.

Table 2. Summary of selected oxychalcogenides with their material properties including thickness (d), conductivity (σ), transmittance (T), and optical band gap (E_g).

Oxychalcogenides	Fabrication	d (nm)	σ (S·cm^{-1})	T (%)	E_g (eV)	Ref.
LaCuOS	RFMS	200	1.2×10^{-2}	70	3.1	[168]
LaCuOS:Sr	RFMS	150	20	60	3.1	[179]
PrCuOS:Sr	RFMS	150	1.8	–	3.03	[174]
NdCuOS:Sr	RFMS	150	0.32	–	2.98	[174]
LaCuOSe:Mg	PLD	40	910	30	2.8	[63]
LaCuOSe	PLD	150	24	–	2.8	[175]
YCuOSe (Bulk)	–	–	0.14	–	2.58	[178]
[Cu$_2$S$_2$][Sr$_3$Sc$_2$O$_5$] (Bulk)	–	–	2.8	–	3.1	[172]
NdCuOS:Mg	Dip coating	200	52.1	54.3	2.91	[189]

YCuOSe was synthesized by Ueda et al. in 2007, and its optical and electrical properties were studied, and this lead to a deeper understanding of the effects of Ln or Ch substitution [178]. However, YCuOSe has a rather small band gap of 2.58 eV, originating from the short Cu–Cu distances. While it has a much higher intrinsic conductivity than LaCuOSe, its small band gap limits its applications for transparent electronics. However, a similar compound, YCuOS, which is expected to depict a larger band gap, has not been successfully prepared yet, so more works should be devoted to achieve such a phase. Other than YCuOSe, another very promising oxychalcogenide [Cu$_2$S$_2$][Sr$_3$Sc$_2$O$_5$] was proposed and synthesized by Liu et al. in polycrystalline bulk form via solid-state reaction in the same year [172]. It has a similar crystal structure as LnCuOCh, with the [Ln$_2$O$_2$]$^{2+}$ layer being replaced with the [Sr$_3$Sc$_2$O$_5$]$^{2+}$ layer (shown in Figure 7b). The rather ionic perovskite [Sr$_3$Sc$_2$O$_5$] layer serves to preserve the large direct band gap of 3.1 eV. As discussed above for the case of Ln substitution, ionic trivalent ions of a larger size such as La and bivalent ions such as Ba could replace Sr and Sc, respectively, to increase the optical band gap [10]. With the band gap enhancement, the possibility of S substitution by Se to further improve the p-type mobility could be realized without affecting the transparency in visible light range. More importantly, there are two cation sites for doping, Sr^{2+} and Sc^{3+}, to enhance the conductivity compared to the single cation site in other oxychalcogenides. The high hole mobility in this layered oxychalcogenide originated from the large dispersion of VBM [10]. Unfortunately, no subsequent experimental work has been reported by any research group for this promising [Cu$_2$S$_2$][Sr$_3$Sc$_2$O$_5$] layered material. This might partly be due to the difficulty of achieving such a phase via conventional methods, but it is worthwhile to re-investigate such a material and unfold its full potential.

So far, the most popular synthesis method to prepare LnCuOCh is solid-state reaction, which involves the mixing of precursor oxides and sulfides such as Ln_2O_3, Ln, Cu_2O, Cu_2S, and La_2S_3, etc., and sintering at high temperature for a long time. While this is the most straightforward preparation method, some of the starting materials such as Ln_2O_3 and Ln are either very sensitive to air or expensive [190]. This led to the development of an alternative synthesis method. In 2008, Nakachi and Ueda developed a flux method to prepare single-crystal LaCuOS in a relatively large size of $3.0 \times 2.8 \times 0.049$ mm^3 with a low conductivity of 7.1×10^{-4} S·cm^{-1} and transmittance around 60% [191]. However, this method has the drawbacks of complexity in the experimental set-up, and long synthesis duration. Alternatively, chemical solution-based solvothermal and precipitation/reduction methods were proposed in 2010 and 2016, respectively, to prepare bulk LaCuOS [192,193]. Single-phase LaCuOS could be produced, and the solvothermal method could synthesize LaCuOS at a relatively low temperature of 200 °C, making it compatible with a flexible substrate such as polyethylene terephthalate (PET). Unfortunately, the important electrical properties of such a material were not characterized in those two reports. Then, in 2017, Zhang et al. proposed a novel two-step synthesis method by combining a solid-state reaction with a sulfurization method [194]. Relatively stable and low-cost starting materials including Cu_2O and La_2O_3 were used for the first step of the solid-state reaction to prepare $CuLaO_2$, and then, the oxide $CuLaO_2$ was sulfurized with S instead of the usually used compound H_2S, which is highly toxic. Single-phase LaCuOS with a relatively high resistivity of 0.25 MΩ was produced. Besides the bulk LnCuOCh synthesis, its thin film form was normally prepared by a vacuum-based deposition method including RFMS, and PLD of the LnCuOCh target prepared via a solid solution method [168,175,181,195]. Such a conventional thin-film fabrication method has the drawback of time-consuming target preparation and the necessity of the use of H_2S during film deposition for both PLD and sputtering. Later, in 2018, Zhang et al. successfully prepared a LaCuOS film via a hydrogen-free method without using H_2S, and the schematic illustration of the fabrication process is shown in Figure 9 [196]. More importantly, the hydrogen-containing gas H_2S was normally used in the thin film deposition process prior to that work. This might impede the p-type performance of oxychalcogenides such as H, which is known to be an n-type dopant, and might be accidentally incorporated in the deposited thin film. Therefore, the absence of H_2S in that work not only serves to be more environmental friendly, but also improves the p-type conductivity, and the intrinsic conductivity in that work is as high as 0.3 S·cm^{-1} [196].

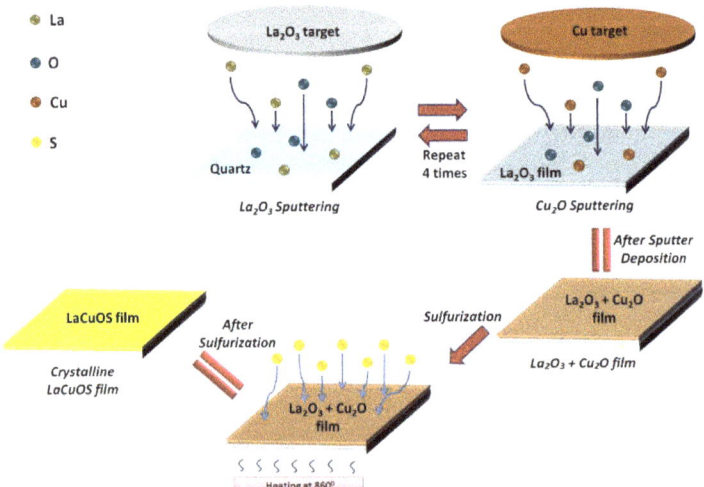

Figure 9. Schematic illustration of the LaCuOS thin film deposition process via sputtering and post-sulfurization. Reproduced with permission from Ref. [196]. Copyright 2018 Elsevier.

Similar to the fabrication of delafossite thin films, the chemical solution method has the advantages of low cost fabrication, roll-to-roll capability, high throughput, and the possibility for a special nanostructure [108–110]. To explore the possibility of a solution-based fabrication approach, in 2018, Zhang et al. developed the first solution-based method, which is a dip-coating process with a post-sulfurization process to fabricate oxychalcogenide NdCuOS (shown in Figure 10) [189]. As no H-containing compound such as H_2S was used during sulfurization, the possibility of hydrogen doping in the thin film was eliminated. Moreover, the NdCuOS prepared via such a solution method was reported to contain a large Cu deficiency. The resulting intrinsic NdCuOS film demonstrated a high p-type conductivity of 6.4 S·cm^{-1}, and even a higher conductivity of 52.1 S·cm^{-1} was achieved after Mg doping with an acceptable optical transparency higher than 50%. Moreover, a working p-n junction was made by using this intrinsic NdCuOS as a p-type terminal, and ZnO doped with Al as an n-type terminal. A low turn-on voltage of 1.1 V and a rather low leakage current of 9.12 µA at −3 V was obtained for this transparent diode [189].

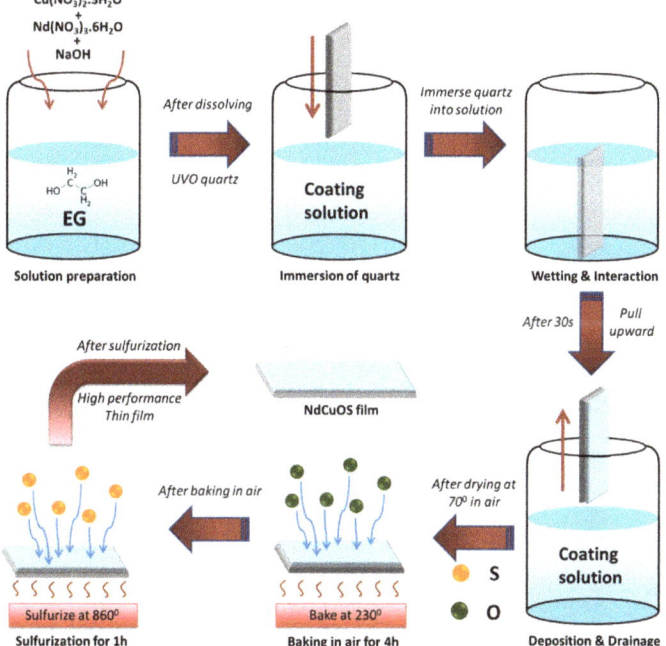

Figure 10. Schematic illustration of the solution-based (dip-coating) process with post-annealing to prepare NdCuOS film. Reproduced with permission from Ref. [189]. Copyright 2018 Elsevier.

2.3. Copper Oxides

Copper oxides have two forms: cupric oxide (CuO) and cuprous oxide (Cu_2O). CuO has a tenorite structure, and Cu_2O has a cubic crystal structure (shown in Figure 11). The optical band gap of CuO is between 1.9–2.1 eV, and that of Cu_2O is between 2.1–2.6 eV, and the low band gaps makes the film non-transparent for a range of the visible light, as the visible light energy reaches 3.1 eV [197,198]. They are both reported to be p-type oxides, but the small band gap limits their applications. Thin-film transistor (TFT) is an important application for copper oxide as an active material. Compared with CuO, Cu_2O can theoretically exhibit a hole mobility exceeding 100 cm^2·V^{-1}·s^{-1}, and so it receives more attention [197]. It is recently accepted that the high p-type mobility of Cu_2O is the result of the band structure of VBM [197]. It has been shown that the p-type conductivity of Cu_2O is from the Cu vacancies [197]. At the same time, it has also been reported that the presence of oxygen interstitials could contribute to p-type conductivity [9,199]. Cu_2O films can be fabricated by various techniques, which are mainly vacuum-based, such as RFMS [64,200–202], ALD [203–205], and PLD [206–208]. In 2008, Li et al. reported a high Hall mobility of 256 $cm^2 \cdot V^{-1} \cdot s^{-1}$ in their Cu_2O thin film deposited by RFMS at a substrate temperature of 600 °C, which is the highest so far for Cu_2O films [64]. Zou et al. reported a pure phase polycrystalline Cu_2O by PLD at 500 °C, with a Hall mobility of 107 $cm^2 \cdot V^{-1} \cdot s^{-1}$ [209]. Jeong and Aydil reported a Cu_2O film with a Hall mobility over 30 $cm^2 \cdot V^{-1} \cdot s^{-1}$ by metal–organic chemical vapor deposition (MOCVD) at 400 °C [210]. Kwon et al. fabricated Cu_2O films by atomic layer deposition with a Hall mobility of 37 $cm^2 \cdot V^{-1} \cdot s^{-1}$ [211]. There are limited reports on the solution-fabricated cuprous oxide without vacuum, and its performance is far from satisfactory. Sun et al., Kim et al., and Yu et al. reported sol–gel methods to fabricate Cu_2O thin films with a Hall mobility of 8, 18.9, and 31.7 $cm^2 \cdot V^{-1} \cdot s^{-1}$, respectively [212–214].

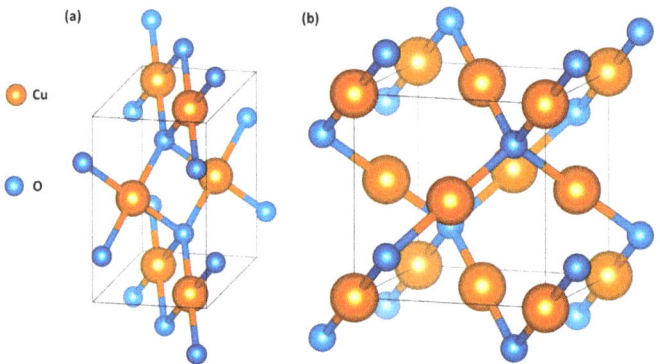

Figure 11. Crystal structure of (**a**) CuO and (**b**) Cu_2O.

In 2008, Matsuzaki et al. reported the first p-type Cu_2O thin film and TFT device by PLD with a Hall mobility of 90 $cm^2 \cdot V^{-1} \cdot s^{-1}$ and field-effect mobility of 0.26 $cm^2 \cdot V^{-1} \cdot s^{-1}$ [215]. Zou et al. also fabricated a p-type Cu_2O film by PLD and obtained a Hall mobility of 107 $cm^2 \cdot V^{-1} \cdot s^{-1}$, and the according TFT shows a field-effect mobility of 4.3 $cm^2 \cdot V^{-1} \cdot s^{-1}$ and an on/off ratio of 10^6, which is the best until now for the copper oxide-based TFTs [209]. Yao et al. reported p-type Cu_2O TFTs by sputtering, which showed a field-effect mobility of 2.4 $cm^2 \cdot V^{-1} \cdot s^{-1}$ [216]. In 2013, Kim et al. presented the first solution processed p-type Cu_2O TFTs with a field-effect mobility of 0.16 $cm^2 \cdot V^{-1} \cdot s^{-1}$ [213]. Anthopoulos et al. reported Cu_2O films by solution-based spray pyrolysis and incorporated them into p-type TFTs with a mobility of 0.01 $cm^2 \cdot V^{-1} \cdot s^{-1}$ and an on/off current ratio of 10^3 [217]. In 2016, Shan et al. fabricated Cu-based oxide TFTs at low-temperature by a solution-processed method with a mobility of 0.78 $cm^2 \cdot V^{-1} \cdot s^{-1}$ and an on/off current ratio of 10^5 [218]. In 2017, Liu et al. reported low-temperature solution-processed p-type Cu_xO thin film-based TFTs with Al_2O_3 dielectrics, which presented a hole mobility of 2.7 $cm^2 \cdot V^{-1} \cdot s^{-1}$ and an on/off ratio of 10^5 [219]. While the device performance is encouraging, it should be noted that the rather low band gap or optical transparency would make Cu_2O have limited applications in transparent devices.

3. Summary and Perspectives

In this review, we have illustrated and discussed the material performance of Cu-based oxides (delafossites and copper oxides) and oxychalcogenides (LnCuOCh) for p-type TCO applications. All three types of materials make use of the special feature of Cu^+ 3d orbitals, which are close to O 2p orbitals and could hybridize with each other to form a more spatially dispersed VBM, leading to higher hole mobility. Moreover, oxychalcogenides LnCuOS VBM is actually formed by the hybridization of Cu 3d orbitals and S 3p orbital states, and this allows for even higher mobility. Over the years, many strategies had been developed to improve the hole mobility of the materials. For example, non-isovalent CuO has been introduced into polycrystalline $CuAlO_2$ film to achieve the hybridization of Cu–O dimers in the $CuAlO_2$ lattice, and resulting in a higher Hall mobility [120]. At the same time, many approaches have been investigated to enhance the p-type conductivity by increasing the hole concentration. Different strategies are suitable for different applications, as certain applications such as TFT would prefer TCO with high mobility, while other applications such as HTL in photovoltaics need TCO with high hole concentrations and conductivity.

While many promising results have been generated recently in terms of the p-type conductivity of Cu-based oxides and oxychalcogenides, their mobility and conductivity are still far poorer than their n-type counterparts. In order to further improve their performance, several important issues could be tackled. Firstly, more high-throughput computational studies could be carried out to find potential p-type TCOs with intrinsically higher hole mobility and the ease of p-type doping. Secondly, both cation and anion doping should be investigated in detail, as only very limited studies have been reported for anion doping for Cu-based oxides and oxychalcogenides. Thirdly, more research efforts should be involved in the study of some reported Cu-based oxides and oxychalcogenides with high potential. For instance, $[Cu_2S_2][Sr_3Sc_2O_5]$ has very high undoped p-type conductivity with two cation sites for doping to further improve both transparency and conductivity, but only one experimental work has been reported for this material. Last but not least, the investigation of bipolar semiconductors should be given more research attention, so that the transparent electronics based on homojunction could be realized. With these four non-exclusive avenues for future p-type TCOs research, we believe that more promising results should be expected.

Funding: This research was funded by Singapore Ministry of Education Academic Research Fund Tier 2 MOE2016-T2-1-049, grant R-284-000-157-112 is appreciated.

Conflicts of Interest: The authors have no conflict of interests.

References

1. Lewis, B.G.; Paine, D.C. Applications and processing of transparent conducting oxides. *MRS Bull.* **2000**, *25*, 22–27. [CrossRef]
2. Fortunato, E.; Barquinha, P.; Martins, R. Oxide semiconductor thin-film transistors: A review of recent advances. *Adv. Mater.* **2012**, *24*, 2945–2986. [CrossRef] [PubMed]
3. Fortunato, E.M.; Barquinha, P.M.; Pimentel, A.; Goncalves, A.M.; Marques, A.J.; Pereira, L.M.; Martins, R.F. Fully transparent zno thin-film transistor produced at room temperature. *Adv. Mater.* **2005**, *17*, 590–594. [CrossRef]
4. Hoel, C.A.; Mason, T.O.; Gaillard, J.-F.; Poeppelmeier, K.R. Transparent conducting oxides in the $ZnO-In_2O_3-SnO_2$ system. *Chem. Mater.* **2010**, *22*, 3569–3579. [CrossRef]
5. Granqvist, C.G.; Hultåker, A. Transparent and conducting ito films: New developments and applications. *Thin Solid Films* **2002**, *411*, 1–5. [CrossRef]
6. Niu, C. Carbon nanotube transparent conducting films. *MRS Bull.* **2011**, *36*, 766–773. [CrossRef]
7. Chen, Z.; Li, W.; Li, R.; Zhang, Y.; Xu, G.; Cheng, H. Fabrication of highly transparent and conductive indium–tin oxide thin films with a high figure of merit via solution processing. *Langmuir* **2013**, *29*, 13836–13842. [CrossRef]
8. Tripathi, T.S.; Karppinen, M. Atomic layer deposition of p-type semiconducting thin films: A review. *Adv. Mater. Interfaces* **2017**, *4*, 1700300. [CrossRef]
9. Raebiger, H.; Lany, S.; Zunger, A. Origins of the p-type nature and cation deficiency in Cu_2O and related materials. *Phys. Rev. B* **2007**, *76*, 045209. [CrossRef]
10. Scanlon, D.O.; Watson, G.W. $(Cu_2S_2)(Sr_3Sc_2O_5)$—A layered, direct band gap, p-type transparent conducting oxychalcogenide: A theoretical analysis. *Chem. Mater.* **2009**, *21*, 5435–5442. [CrossRef]
11. Banerjee, A.; Chattopadhyay, K. Recent developments in the emerging field of crystalline p-type transparent conducting oxide thin films. *Prog. Cryst. Growth Charact. Mater.* **2005**, *50*, 52–105. [CrossRef]

12. Hautier, G.; Miglio, A.; Ceder, G.; Rignanese, G.-M.; Gonze, X. Identification and design principles of low hole effective mass p-type transparent conducting oxides. *Nat. Commun.* **2013**, *4*, 2292. [CrossRef] [PubMed]
13. Hiramatsu, H.; Kamiya, T.; Tohei, T.; Ikenaga, E.; Mizoguchi, T.; Ikuhara, Y.; Kobayashi, K.; Hosono, H. Origins of hole doping and relevant optoelectronic properties of wide gap p-type semiconductor, lacuose. *J. Am. Chem. Soc.* **2010**, *132*, 15060–15067. [CrossRef] [PubMed]
14. Zakutayev, A.; McIntyre, D.; Schneider, G.; Kykyneshi, R.; Keszler, D.; Park, C.-H.; Tate, J. Tunable properties of wide-band gap p-type bacu ($Ch_{1-x}Ch_x'$)F (Ch = S, Se, Te) thin-film solid solutions. *Thin Solid Films* **2010**, *518*, 5494–5500. [CrossRef]
15. Wijeyasinghe, N.; Regoutz, A.; Eisner, F.; Du, T.; Tsetseris, L.; Lin, Y.H.; Faber, H.; Pattanasattayavong, P.; Li, J.; Yan, F. Copper(I) thiocyanate (CuSCN) hole-transport layers processed from aqueous precursor solutions and their application in thin-film transistors and highly efficient organic and organometal halide perovskite solar cells. *Adv. Funct. Mater.* **2017**, *27*, 1701818. [CrossRef]
16. Liu, X.; Zhang, N.; Tang, B.; Li, M.; Zhang, Y.-W.; Yu, Z.G.; Gong, H. Highly stable new organic–inorganic hybrid 3D perovskite $CH_3NH_3PdI_3$ and 2D perovskite $(CH_3NH_3)_3Pd_2I_7$: DFT analysis, synthesis, structure, transition behavior, and physical properties. *J. Phys. Chem. Lett.* **2018**, *9*, 5862–5872. [CrossRef] [PubMed]
17. Liu, X.; Li, B.; Zhang, N.; Yu, Z.; Sun, K.; Tang, B.; Shi, D.; Yao, H.; Ouyang, J.; Gong, H. Multifunctional RbCl dopants for efficient inverted planar perovskite solar cell with ultra-high fill factor, negligible hysteresis and improved stability. *Nano Energy* **2018**, *53*, 567–578. [CrossRef]
18. Liu, X.; Huang, T.J.; Zhang, L.; Tang, B.; Zhang, N.; Shi, D.; Gong, H. Highly stable, new, organic-inorganic perovskite $(CH_3NH_3)_2PdBr_4$: Synthesis, structure, and physical properties. *Chem. Eur. J.* **2018**, *24*, 4991–4998. [CrossRef] [PubMed]
19. Klauk, H.; Zschieschang, U.; Pflaum, J.; Halik, M. Ultralow-power organic complementary circuits. *Nature* **2007**, *445*, 745. [CrossRef]
20. Martins, R.; Nathan, A.; Barros, R.; Pereira, L.; Barquinha, P.; Correia, N.; Costa, R.; Ahnood, A.; Ferreira, I.; Fortunato, E. Complementary metal oxide semiconductor technology with and on paper. *Adv. Mater.* **2011**, *23*, 4491–4496. [CrossRef]
21. Martins, R.F.; Ahnood, A.; Correia, N.; Pereira, L.M.; Barros, R.; Barquinha, P.M.; Costa, R.; Ferreira, I.M.; Nathan, A.; Fortunato, E.E. Recyclable, flexible, low-power oxide electronics. *Adv. Funct. Mater.* **2013**, *23*, 2153–2161. [CrossRef]
22. Wager, J.F.; Hoffman, R. Thin, fast, and flexible. *IEEE Spectr.* **2011**, *48*, 42–56. [CrossRef]
23. Kawazoe, H.; Yanagi, H.; Ueda, K.; Hosono, H. Transparent p-type conducting oxides: Design and fabrication of pn heterojunctions. *MRS Bull.* **2000**, *25*, 28–36. [CrossRef]
24. Kawazoe, H.; Yasukawa, M.; Hyodo, H.; Kurita, M.; Yanagi, H.; Hosono, H. p-type electrical conduction in transparent thin films of $CuAlO_2$. *Nature* **1997**, *389*, 939. [CrossRef]
25. Hu, Z.; Huang, X.; Annapureddy, H.V.; Margulis, C.J. Molecular dynamics study of the temperature-dependent optical kerr effect spectra and intermolecular dynamics of room temperature ionic liquid 1-methoxyethylpyridinium dicyanoamide. *J. Phys. Chem. B* **2008**, *112*, 7837–7849. [CrossRef] [PubMed]
26. Gong, H.; Wang, Y.; Luo, Y. Nanocrystalline p-type transparent Cu–Al–O semiconductor prepared by chemical-vapor deposition with $Cu(acac)_2$ and $Al(acac)_3$ precursors. *Appl. Phys. Lett.* **2000**, *76*, 3959–3961. [CrossRef]
27. Ohashi, M.; Iida, Y.; Morikawa, H. Preparation of $CuAlO_2$ films by wet chemical synthesis. *J. Am. Ceram. Soc.* **2002**, *85*, 270–272. [CrossRef]
28. Banerjee, A.; Kundoo, S.; Chattopadhyay, K. Synthesis and characterization of p-type transparent conducting $CuAlO_2$ thin film by dc sputtering. *Thin Solid Films* **2003**, *440*, 5–10. [CrossRef]
29. Mahapatra, S.; Shivashankar, S.A. Low-pressure metal–organic chemical vapor deposition of transparent and p-type conducting $CuCrO_2$ thin films with high conductivity. *Chem. Vap. Depos.* **2003**, *9*, 238–240. [CrossRef]
30. Nagarajan, R.; Draeseke, A.; Sleight, A.; Tate, J. p-Type conductivity in $CuCr_{1-x}Mg_xO_2$ films and powders. *J. Appl. Phys.* **2001**, *89*, 8022–8025. [CrossRef]
31. Bywalez, R.; Götzendörfer, S.; Löbmann, P. Structural and physical effects of Mg-doping on p-type $CuCrO_2$ and $CuAl_{0.5}Cr_{0.5}O_2$ thin films. *J. Mater. Chem.* **2010**, *20*, 6562–6570. [CrossRef]
32. Ueda, K.; Hase, T.; Yanagi, H.; Kawazoe, H.; Hosono, H.; Ohta, H.; Orita, M.; Hirano, M. Epitaxial growth of transparent p-type conducting $CuGaO_2$ thin films on sapphire (001) substrates by pulsed laser deposition. *J. Appl. Phys.* **2001**, *89*, 1790–1793. [CrossRef]

33. Yanagi, H.; Kawazoe, H.; Kudo, A.; Yasukawa, M.; Hosono, H. Chemical design and thin film preparation of p-type conductive transparent oxides. *J. Electroceram.* **2000**, *4*, 407–414. [CrossRef]
34. Srinivasan, R.; Chavillon, B.; Doussier-Brochard, C.; Cario, L.; Paris, M.; Gautron, E.; Deniard, P.; Odobel, F.; Jobic, S. Tuning the size and color of the p-type wide band gap delafossite semiconductor $CuGaO_2$ with ethylene glycol assisted hydrothermal synthesis. *J. Mater. Chem.* **2008**, *18*, 5647–5653. [CrossRef]
35. Yanagi, H.; Hase, T.; Ibuki, S.; Ueda, K.; Hosono, H. Bipolarity in electrical conduction of transparent oxide semiconductor $CuInO_2$ with delafossite structure. *Appl. Phys. Lett.* **2001**, *78*, 1583–1585. [CrossRef]
36. Sasaki, M.; Shimode, M. Fabrication of bipolar $CuInO_2$ with delafossite structure. *J. Phys. Chem. Solids* **2003**, *64*, 1675–1679. [CrossRef]
37. Yanagi, H.; Ueda, K.; Ohta, H.; Orita, M.; Hirano, M.; Hosono, H. Fabrication of all oxide transparent p–n homojunction using bipolar $CuInO_2$ semiconducting oxide with delafossite structure. *Solid State Commun.* **2001**, *121*, 15–17. [CrossRef]
38. Duan, N.; Sleight, A.; Jayaraj, M.; Tate, J. Transparent p-type conducting $CuScO_{2+x}$ films. *Appl. Phys. Lett.* **2000**, *77*, 1325–1326. [CrossRef]
39. Kakehi, Y.; Nakao, S.; Satoh, K.; Yotsuya, T. Properties of copper-scandium oxide thin films prepared by pulsed laser deposition. *Thin Solid Films* **2003**, *445*, 294–298. [CrossRef]
40. Kakehi, Y.; Satoh, K.; Yotsuya, T.; Nakao, S.; Yoshimura, T.; Ashida, A.; Fujimura, N. Epitaxial growth of $CuScO_2$ thin films on sapphire a-plane substrates by pulsed laser deposition. *J. Appl. Phys.* **2005**, *97*, 083535. [CrossRef]
41. Manoj, R.; Nisha, M.; Vanaja, K.; Jayaraj, M. Effect of oxygen intercalation on properties of sputtered $CuYO_2$ for potential use as p-type transparent conducting films. *Bull. Mater. Sci.* **2008**, *31*, 49–53. [CrossRef]
42. Nagarajan, R.; Duan, N.; Jayaraj, M.; Li, J.; Vanaja, K.; Yokochi, A.; Draeseke, A.; Tate, J.; Sleight, A. p-type conductivity in the delafossite structure. *Int. J. Inorg. Mater.* **2001**, *3*, 265–270. [CrossRef]
43. Ingram, B.J.; González, G.B.; Mason, T.O.; Shahriari, D.Y.; Barnabe, A.; Ko, D.; Poeppelmeier, K.R. Transport and defect mechanisms in cuprous delafossites. 1. Comparison of hydrothermal and standard solid-state synthesis in $CuAlO_2$. *Chem. Mater.* **2004**, *16*, 5616–5622. [CrossRef]
44. Chen, H.-Y.; Wu, J.-H. Characterization and optoelectronic properties of sol–gel-derived $CuFeO_2$ thin films. *Thin Solid Films* **2012**, *520*, 5029–5035. [CrossRef]
45. Chiu, T.-W.; Huang, P.-S. Preparation of delafossite $CuFeO_2$ coral-like powder using a self-combustion glycine nitrate process. *Ceram. Int.* **2013**, *39*, S575–S578. [CrossRef]
46. Moharam, M.; Rashad, M.; Elsayed, E.; Abou-Shahba, R. A facile novel synthesis of delafossite $CuFeO_2$ powders. *J. Mater. Sci. Mater. Electron.* **2014**, *25*, 1798–1803. [CrossRef]
47. Snure, M.; Tiwari, A. $CuBO_2$: A p-type transparent oxide. *Appl. Phys. Lett.* **2007**, *91*, 092123. [CrossRef]
48. Santra, S.; Das, N.; Chattopadhyay, K. Physical and optical properties of $CuBO_2$ nanopowders synthesized via sol-gel route. In Proceedings of the 16th International Workshop on Physics of Semiconductor Devices, Kanpur, India, 19–22 December 2011; p. 85491W.
49. Iozzi, M.; Vajeeston, P.; Vidya, R.; Ravindran, P.; Fjellvåg, H. Structural and electronic properties of transparent conducting delafossite: A comparison between the $AgBO_2$ and $CuBO_2$ families (B = Al, Ga, In and Sc, Y). *RSC Adv.* **2015**, *5*, 1366–1377. [CrossRef]
50. Kudo, A.; Yanagi, H.; Hosono, H.; Kawazoe, H. $SrCu_2O_2$: A p-type conductive oxide with wide band gap. *Appl. Phys. Lett.* **1998**, *73*, 220–222. [CrossRef]
51. Ohta, H.; Kawamura, K.-I.; Orita, M.; Hirano, M.; Sarukura, N.; Hosono, H. Current injection emission from a transparent p–n junction composed of p-$SrCu_2O_2$/n-ZnO. *Appl. Phys. Lett.* **2000**, *77*, 475–477. [CrossRef]
52. Ohta, H.; Orita, M.; Hirano, M.; Hosono, H. Fabrication and characterization of ultraviolet-emitting diodes composed of transparent p–n heterojunction, p-$SrCu_2O_2$ and n-ZnO. *J. Appl. Phys.* **2001**, *89*, 5720–5725. [CrossRef]
53. Kim, J.; Yamamoto, K.; Iimura, S.; Ueda, S.; Hosono, H. Electron affinity control of amorphous oxide semiconductors and its applicability to organic electronics. *Adv. Mater. Interfaces* **2018**, *5*, 1801307. [CrossRef]
54. Zhang, H.; Wang, H.; Chen, W.; Jen, A.K.Y. $CuGaO_2$: A promising inorganic hole-transporting material for highly efficient and stable perovskite solar cells. *Adv. Mater.* **2017**, *29*, 1604984. [CrossRef] [PubMed]
55. Mryasov, O.; Freeman, A. Electronic band structure of indium tin oxide and criteria for transparent conducting behavior. *Phys. Rev. B* **2001**, *64*, 233111. [CrossRef]

56. Scanlon, D.O.; Watson, G.W. Understanding the p-type defect chemistry of $CuCrO_2$. *J. Mater. Chem.* **2011**, *21*, 3655–3663. [CrossRef]
57. Ingram, B.; Mason, T.; Asahi, R.; Park, K.; Freeman, A. Electronic structure and small polaron hole transport of copper aluminate. *Phys. Rev. B* **2001**, *64*, 155114. [CrossRef]
58. Shin, D.; Foord, J.; Egdell, R.; Walsh, A. Electronic structure of $CuCrO_2$ thin films grown on Al_2O_3 (001) by oxygen plasma assisted molecular beam epitaxy. *J. Appl. Phys.* **2012**, *112*, 113718. [CrossRef]
59. Zhang, Y.; Qiao, Z.-P.; Chen, X.-M. Microwave-assisted elemental direct reaction route to nanocrystalline copper chalcogenides cuse and Cu_2Te. *J. Mater. Chem.* **2002**, *12*, 2747–2748. [CrossRef]
60. Liu, M.L.; Wu, L.B.; Huang, F.Q.; Chen, L.D.; Ibers, J.A. Syntheses, crystal and electronic structure, and some optical and transport properties of lncuote (Ln = La, Ce, Nd). *J. Solid State Chem.* **2007**, *180*, 62–69. [CrossRef]
61. Ueda, K.; Inoue, S.; Hosono, H.; Sarukura, N.; Hirano, M. Room-temperature excitons in wide-gap layered-oxysulfide semiconductor: Lacuos. *Appl. Phys. Lett.* **2001**, *78*, 2333–2335. [CrossRef]
62. Ueda, K.; Hiramatsu, H.; Ohta, H.; Hirano, M.; Kamiya, T.; Hosono, H. Single-atomic-layered quantum wells built in wide-gap semiconductors LnCuOCh (Ln = lanthanide, Ch = chalcogen). *Phys. Rev. B* **2004**, *69*, 155305. [CrossRef]
63. Hiramatsu, H.; Ueda, K.; Ohta, H.; Hirano, M.; Kikuchi, M.; Yanagi, H.; Kamiya, T.; Hosono, H. Heavy hole doping of epitaxial thin films of a wide gap p-type semiconductor, lacuose, and analysis of the effective mass. *Appl. Phys. Lett.* **2007**, *91*, 012104. [CrossRef]
64. Li, B.; Akimoto, K.; Shen, A. Growth of Cu_2O thin films with high hole mobility by introducing a low-temperature buffer layer. *J. Cryst. Growth* **2009**, *311*, 1102–1105. [CrossRef]
65. De Jongh, P.; Vanmaekelbergh, D.; Kelly, J. Cu_2O: Electrodeposition and characterization. *Chem. Mater.* **1999**, *11*, 3512–3517. [CrossRef]
66. Wang, Y.; Miska, P.; Pilloud, D.; Horwat, D.; Mücklich, F.; Pierson, J.-F. Transmittance enhancement and optical band gap widening of Cu_2O thin films after air annealing. *J. Appl. Phys.* **2014**, *115*, 073505. [CrossRef]
67. Nian, J.-N.; Hu, C.-C.; Teng, H. Electrodeposited p-type Cu_2O for H_2 evolution from photoelectrolysis of water under visible light illumination. *Int. J. Hydrog. Energy* **2008**, *33*, 2897–2903. [CrossRef]
68. Ruiz, E.; Alvarez, S.; Alemany, P.; Evarestov, R.A. Electronic structure and properties of Cu_2O. *Phys. Rev. B* **1997**, *56*, 7189. [CrossRef]
69. Scanlon, D.O.; Morgan, B.J.; Watson, G.W.; Walsh, A. Acceptor levels in p-type Cu_2O: Rationalizing theory and experiment. *Phys. Rev. Lett.* **2009**, *103*, 096405. [CrossRef]
70. Ishiguro, T.; Kitazawa, A.; Mizutani, N.; Kato, M. Single-crystal growth and crystal structure refinement of $CuAlO_2$. *J. Solid State Chem.* **1981**, *40*, 170–174. [CrossRef]
71. Marquardt, M.A.; Ashmore, N.A.; Cann, D.P. Crystal chemistry and electrical properties of the delafossite structure. *Thin Solid Films* **2006**, *496*, 146–156. [CrossRef]
72. Wei, R.; Tang, X.; Hu, L.; Yang, J.; Zhu, X.; Song, W.; Dai, J.; Zhu, X.; Sun, Y. Facile chemical solution synthesis of p-type delafossite Ag-based transparent conducting $AgCrO_2$ films in an open condition. *J. Mater. Chem. C* **2017**, *5*, 1885–1892. [CrossRef]
73. Benko, F.; Koffyberg, F. The optical interband transitions of the semiconductor $CuGaO_2$. *Physica Status Solidi (a)* **1986**, *94*, 231–234. [CrossRef]
74. Vanaja, K.; Ajimsha, R.; Asha, A.; Jayaraj, M. p-type electrical conduction in α-$AgGaO_2$ delafossite thin films. *Appl. Phys. Lett.* **2006**, *88*, 212103. [CrossRef]
75. Kandpal, H.C.; Seshadri, R. First-principles electronic structure of the delafossites ABO_2 (A= Cu, Ag, Au; B= Al, Ga, Sc, In, Y): Evolution of d^{10}–d^{10} interactions. *Solid State Sci.* **2002**, *4*, 1045–1052. [CrossRef]
76. Huda, M.N.; Yan, Y.; Walsh, A.; Wei, S.-H.; Al-Jassim, M.M. Group-IIIA versus IIIB delafossites: Electronic structure study. *Phys. Rev. B* **2009**, *80*, 035205. [CrossRef]
77. Pellicer-Porres, J.; Segura, A.; Gilliland, A.; Munoz, A.; Rodríguez-Hernández, P.; Kim, D.; Lee, M.; Kim, T. On the band gap of $CuAlO_2$ delafossite. *Appl. Phys. Lett.* **2006**, *88*, 181904. [CrossRef]
78. Gilliland, S.; Pellicer-Porres, J.; Segura, A.; Muñoz, A.; Rodríguez-Hernández, P.; Kim, D.; Lee, M.; Kim, T. Electronic structure of $CuAlO_2$ and $CuScO_2$ delafossites under pressure. *Phys. Status Solidi (b)* **2007**, *244*, 309–314. [CrossRef]
79. Jayalakshmi, V.; Murugan, R.; Palanivel, B. Electronic and structural properties of $CuMO_2$ (M = Al, Ga, In). *J. Alloy. Compd.* **2005**, *388*, 19–22. [CrossRef]

80. Buljan, A.; Alemany, P.; Ruiz, E. Electronic structure and bonding in CuMO$_2$ (M = Al, Ga, Y) delafossite-type oxides: An ab initio study. *J. Phys. Chem. B* **1999**, *103*, 8060–8066. [CrossRef]
81. Schiavo, E.; Latouche, C.; Barone, V.; Crescenzi, O.; Muñoz-García, A.B.; Pavone, M. An ab initio study of Cu-based delafossites as an alternative to nickel oxide in photocathodes: Effects of Mg-doping and surface electronic features. *Phys. Chem. Chem. Phys.* **2018**, *20*, 14082–14089. [CrossRef] [PubMed]
82. Sakulkalavek, A.; Sakdanuphab, R. Power factor improvement of delafossite CuAlO$_2$ by liquid-phase sintering with Ag$_2$O addition. *Mater. Sci. Semicond. Process.* **2016**, *56*, 313–323. [CrossRef]
83. Li, J.; Sleight, A.; Jones, C.; Toby, B. Trends in negative thermal expansion behavior for AMO2 (A = Cu Or Ag; M = Al, Sc, In, or La) compounds with the delafossite structure. *J. Solid State Chem.* **2005**, *178*, 285–294. [CrossRef]
84. Tsuboi, N.; Hoshino, T.; Ohara, H.; Suzuki, T.; Kobayashi, S.; Kato, K.; Kaneko, F. Control of luminescence and conductivity of delafossite-type CuYO$_2$ by substitution of rare earth cation (Eu, Tb) and/or Ca cation for Y cation. *J. Phys. Chem. Solids* **2005**, *66*, 2134–2138. [CrossRef]
85. Ruttanapun, C. Optical and electronic properties of delafossite CuBO$_2$ p-type transparent conducting oxide. *J. Appl. Phys.* **2013**, *114*, 113108. [CrossRef]
86. Scanlon, D.O.; Walsh, A.; Watson, G.W. Understanding the p-type conduction properties of the transparent conducting oxide CuBO$_2$: A density functional theory analysis. *Chem. Mater.* **2009**, *21*, 4568–4576. [CrossRef]
87. Scanlon, D.O.; Watson, G.W. Conductivity limits in CuAlO$_2$ from screened-hybrid density functional theory. *J. Phys. Chem. Lett.* **2010**, *1*, 3195–3199. [CrossRef]
88. Benko, F.; Koffyberg, F. Opto-electronic properties of CuAlO$_2$. *J. Phys. Chem. Solids* **1984**, *45*, 57–59. [CrossRef]
89. Fang, M.; He, H.; Lu, B.; Zhang, W.; Zhao, B.; Ye, Z.; Huang, J. Optical properties of p-type CuAlO$_2$ thin film grown by rf magnetron sputtering. *Appl. Surf. Sci.* **2011**, *257*, 8330–8333. [CrossRef]
90. Tsuboi, N.; Takahashi, Y.; Kobayashi, S.; Shimizu, H.; Kato, K.; Kaneko, F. Delafossite CuAlO$_2$ films prepared by reactive sputtering using Cu and Al targets. *J. Phys. Chem. Solids* **2003**, *64*, 1671–1674. [CrossRef]
91. Stauber, R.; Perkins, J.D.; Parilla, P.A.; Ginley, D.S. Thin film growth of transparent p-type CuAlO$_2$. *Electrochem. Solid-State Lett.* **1999**, *2*, 654–656. [CrossRef]
92. Tripathi, T.; Niemelä, J.-P.; Karppinen, M. Atomic layer deposition of transparent semiconducting oxide CuCrO$_2$ thin films. *J. Mater. Chem. C* **2015**, *3*, 8364–8371. [CrossRef]
93. Tripathi, T.S.; Karppinen, M. Enhanced p-type transparent semiconducting characteristics for ALD-grown Mg-substituted CuCrO$_2$ thin films. *Adv. Electron. Mater.* **2017**, *3*, 1600341. [CrossRef]
94. Gao, S.; Zhao, Y.; Gou, P.; Chen, N.; Xie, Y. Preparation of CuAlO$_2$ nanocrystalline transparent thin films with high conductivity. *Nanotechnology* **2003**, *14*, 538. [CrossRef]
95. Sheets, W.C.; Mugnier, E.; Barnabe, A.; Marks, T.J.; Poeppelmeier, K.R. Hydrothermal synthesis of delafossite-type oxides. *Chem. Mater.* **2006**, *18*, 7–20. [CrossRef]
96. Götzendörfer, S.; Polenzky, C.; Ulrich, S.; Löbmann, P. Preparation of CuAlO$_2$ and CuCrO$_2$ thin films by sol–gel processing. *Thin Solid Films* **2009**, *518*, 1153–1156. [CrossRef]
97. Prévot, M.S.; Li, Y.; Guijarro, N.; Sivula, K. Improving charge collection with delafossite photocathodes: A host–guest CuAlO$_2$/CuFeO$_2$ approach. *J. Mater. Chem. A* **2016**, *4*, 3018–3026. [CrossRef]
98. Chiu, S.; Huang, J. Characterization of p-type CuAlO$_2$ thin films grown by chemical solution deposition. *Surf. Coat. Technol.* **2013**, *231*, 239–242. [CrossRef]
99. Phani, P.S.; Vishnukanthan, V.; Sundararajan, G. Effect of heat treatment on properties of cold sprayed nanocrystalline copper alumina coatings. *Acta Mater.* **2007**, *55*, 4741–4751. [CrossRef]
100. Bouzidi, C.; Bouzouita, H.; Timoumi, A.; Rezig, B. Fabrication and characterization of CuAlO$_2$ transparent thin films prepared by spray technique. *Mater. Sci. Eng. B* **2005**, *118*, 259–263. [CrossRef]
101. Grilli, M.; Menchini, F.; Dikonimos, T.; Nunziante, P.; Pilloni, L.; Yilmaz, M.; Piegari, A.; Mittiga, A. Effect of growth parameters on the properties of RF-sputtered highly conductive and transparent p-type NiO$_x$ films. *Semicond. Sci. Technol.* **2016**, *31*, 055016. [CrossRef]
102. Lu, Y.; He, Y.; Yang, B.; Polity, A.; Volbers, N.; Neumann, C.; Hasselkamp, D.; Meyer, B. RF reactive sputter deposition and characterization of transparent CuAlO$_2$ thin films. *Phys. Status Solidi c* **2006**, *3*, 2895–2898. [CrossRef]
103. Zhang, K.H.; Xi, K.; Blamire, M.G.; Egdell, R.G. P-type transparent conducting oxides. *J. Phys. Condens. Matter* **2016**, *28*, 383002. [CrossRef] [PubMed]

104. Xiao, B.; Ye, Z.; Zhang, Y.; Zeng, Y.; Zhu, L.; Zhao, B. Fabrication of p-type Li-doped ZnO films by pulsed laser deposition. *Appl. Surf. Sci.* **2006**, *253*, 895–897. [CrossRef]
105. Ryu, Y.; Kim, W.; White, H. Fabrication of homostructural ZnO p–n junctions. *J. Cryst. Growth* **2000**, *219*, 419–422. [CrossRef]
106. Leskelä, M.; Ritala, M. Atomic layer deposition chemistry: Recent developments and future challenges. *Angew. Chem. Int. Ed.* **2003**, *42*, 5548–5554. [CrossRef] [PubMed]
107. Miikkulainen, V.; Leskelä, M.; Ritala, M.; Puurunen, R.L. Crystallinity of inorganic films grown by atomic layer deposition: Overview and general trends. *J. Appl. Phys.* **2013**, *113*, 2. [CrossRef]
108. Das, B.; Renaud, A.; Volosin, A.M.; Yu, L.; Newman, N.; Seo, D.-K. Nanoporous delafossite $CuAlO_2$ from inorganic/polymer double gels: A desirable high-surface-area p-type transparent electrode material. *Inorg. Chem.* **2015**, *54*, 1100–1108. [CrossRef] [PubMed]
109. Nie, S.; Liu, A.; Meng, Y.; Shin, B.; Liu, G.; Shan, F. Solution-processed ternary p-type $CuCrO_2$ semiconductor thin films and their application in transistors. *J. Mater. Chem. C* **2018**, *6*, 1393–1398. [CrossRef]
110. Ginley, D.; Roy, B.; Ode, A.; Warmsingh, C.; Yoshida, Y.; Parilla, P.; Teplin, C.; Kaydanova, T.; Miedaner, A.; Curtis, C. Non-vacuum and pld growth of next generation tco materials. *Thin Solid Films* **2003**, *445*, 193–198. [CrossRef]
111. Yanagi, H.; Inoue, S.-I.; Ueda, K.; Kawazoe, H.; Hosono, H.; Hamada, N. Electronic structure and optoelectronic properties of transparent p-type conducting $CuAlO_2$. *J. Appl. Phys.* **2000**, *88*, 4159–4163. [CrossRef]
112. Aston, D.; Payne, D.; Green, A.; Egdell, R.; Law, D.; Guo, J.; Glans, P.; Learmonth, T.; Smith, K. High-resolution X-ray spectroscopic study of the electronic structure of the prototypical p-type transparent conducting oxide $CuAlO_2$. *Phys. Rev. B* **2005**, *72*, 195115. [CrossRef]
113. Katayama-Yoshida, H.; Koyanagi, T.; Funashima, H.; Harima, H.; Yanase, A. Engineering of nested fermi surface and transparent conducting p-type delafossite $CuAlO_2$: Possible lattice instability or transparent superconductivity? *Solid State Commun.* **2003**, *126*, 135–139. [CrossRef]
114. Koyanagi, T.; Harima, H.; Yanase, A.; Katayama-Yoshida, H. Materials design of p-type transparent conducting oxides of delafossite $CuAlO_2$ by super-cell FLAPW method. *J. Phys. Chem. Solids* **2003**, *64*, 1443–1446. [CrossRef]
115. Dong, G.; Zhang, M.; Lan, W.; Dong, P.; Yan, H. Structural and physical properties of Mg-doped $CuAlO_2$ thin films. *Vacuum* **2008**, *82*, 1321–1324. [CrossRef]
116. Durá, O.; Boada, R.; Rivera-Calzada, A.; León, C.; Bauer, E.; de la Torre, M.L.; Chaboy, J. Transport, electronic, and structural properties of nanocrystalline $CuAlO_2$ delafossites. *Phys. Rev. B* **2011**, *83*, 045202. [CrossRef]
117. Lee, M.; Kim, T.; Kim, D. Anisotropic electrical conductivity of delafossite-type $CuAlO_2$ laminar crystal. *Appl. Phys. Lett.* **2001**, *79*, 2028–2030. [CrossRef]
118. Tate, J.; Ju, H.; Moon, J.; Zakutayev, A.; Richard, A.; Russell, J.; McIntyre, D. Origin of p-type conduction in single-crystal $CuAlO_2$. *Phys. Rev. B* **2009**, *80*, 165206. [CrossRef]
119. Luo, J.; Lin, Y.-J.; Hung, H.-C.; Liu, C.-J.; Yang, Y.-W. Tuning the formation of p-type defects by peroxidation of $CuAlO_2$ films. *J. Appl. Phys.* **2013**, *114*, 033712.
120. Yao, Z.; He, B.; Zhang, L.; Zhuang, C.; Ng, T.; Liu, S.; Vogel, M.; Kumar, A.; Zhang, W.; Lee, C. Energy band engineering and controlled p-type conductivity of $CuAlO_2$ thin films by nonisovalent Cu–O alloying. *Appl. Phys. Lett.* **2012**, *100*, 062102. [CrossRef]
121. Rastogi, A.; Lim, S.; Desu, S. Structure and optoelectronic properties of spray deposited Mg doped p-$CuCrO_2$ semiconductor oxide thin films. *J. Appl. Phys.* **2008**, *104*, 023712. [CrossRef]
122. Chen, H.-Y.; Yang, C.-C. Transparent p-type Zn-doped $CuCrO_2$ films by sol–gel processing. *Surf. Coat. Technol.* **2013**, *231*, 277–280. [CrossRef]
123. Dong, G.; Zhang, M.; Zhao, X.; Yan, H.; Tian, C.; Ren, Y. Improving the electrical conductivity of $CuCrO_2$ thin film by n doping. *Appl. Surf. Sci.* **2010**, *256*, 4121–4124. [CrossRef]
124. Ahmadi, M.; Asemi, M.; Ghanaatshoar, M. Mg and N co-doped $CuCrO_2$: A record breaking p-type TCO. *Appl. Phys. Lett.* **2018**, *113*, 242101. [CrossRef]
125. Crêpellière, J.; Popa, P.L.; Bahlawane, N.; Leturcq, R.; Werner, F.; Siebentritt, S.; Lenoble, D. Transparent conductive $CuCrO_2$ thin films deposited by pulsed injection metal organic chemical vapor deposition: Up-scalable process technology for an improved transparency/conductivity trade-off. *J. Mater. Chem. C* **2016**, *4*, 4278–4287. [CrossRef]

126. Popa, P.L.; Crepelliere, J.; Nukala, P.; Leturcq, R.; Lenoble, D. Invisible electronics: Metastable Cu-vacancies chain defects for highly conductive p-type transparent oxide. *Appl. Mater. Today* **2017**, *9*, 184–191. [CrossRef]
127. Farrell, L.; Norton, E.; O'Dowd, B.; Caffrey, D.; Shvets, I.; Fleischer, K. Spray pyrolysis growth of a high figure of merit, nano-crystalline, p-type transparent conducting material at low temperature. *Appl. Phys. Lett.* **2015**, *107*, 031901. [CrossRef]
128. Papadopoulou, E.; Viskadourakis, Z.; Pennos, A.; Huyberechts, G.; Aperathitis, E. The effect of deposition parameters on the properties of $SrCu_2O_2$ films fabricated by pulsed laser deposition. *Thin Solid Films* **2008**, *516*, 1449–1452. [CrossRef]
129. Arnold, T.; Payne, D.; Bourlange, A.; Hu, J.; Egdell, R.; Piper, L.; Colakerol, L.; De Masi, A.; Glans, P.-A.; Learmonth, T. X-ray spectroscopic study of the electronic structure of $CuCrO_2$. *Phys. Rev. B* **2009**, *79*, 075102. [CrossRef]
130. Scanlon, D.O.; Walsh, A.; Morgan, B.J.; Watson, G.W.; Payne, D.J.; Egdell, R.G. Effect of cr substitution on the electronic structure of $CuAl_{1-x}Cr_xO_2$. *Phys. Rev. B* **2009**, *79*, 035101. [CrossRef]
131. Scanlon, D.O.; Godinho, K.G.; Morgan, B.J.; Watson, G.W. Understanding conductivity anomalies in Cu^I-based delafossite transparent conducting oxides: Theoretical insights. *J. Chem. Phys.* **2010**, *132*, 024707. [CrossRef]
132. Zheng, S.; Jiang, G.; Su, J.; Zhu, C. The structural and electrical property of $CuCr_{1-x}Ni_xO_2$ delafossite compounds. *Mater. Lett.* **2006**, *60*, 3871–3873. [CrossRef]
133. Kaya, I.C.; Sevindik, M.A.; Akyıldız, H. Characteristics of Fe-and Mg-doped $CuCrO_2$ nanocrystals prepared by hydrothermal synthesis. *J. Mater. Sci. Mater. Electron.* **2016**, *27*, 2404–2411. [CrossRef]
134. Okuda, T.; Jufuku, N.; Hidaka, S.; Terada, N. Magnetic, transport, and thermoelectric properties of the delafossite oxides $CuCr_{1-x}Mg_xO_2$ ($0 \leq x \leq 0.04$). *Phys. Rev. B* **2005**, *72*, 144403. [CrossRef]
135. Madre, M.; Torres, M.; Gomez, J.; Diez, J.; Sotelo, A. Effect of alkaline earth dopant on density, mechanical, and electrical properties of $Cu_{0.97}AE_{0.03}CrO_2$ (AE = Mg, Ca, Sr, and Ba) delafossite oxide. *J. Aust. Ceram. Soc.* **2019**, *55*, 257–263. [CrossRef]
136. Xu, Y.; Nie, G.-Z.; Zou, D.; Tang, J.-W.; Ao, Z. N–Mg dual-acceptor co-doping in $CuCrO_2$ studied by first-principles calculations. *Phys. Lett. A* **2016**, *380*, 3861–3865. [CrossRef]
137. Mandal, P.; Mazumder, N.; Saha, S.; Ghorai, U.K.; Roy, R.; Das, G.C.; Chattopadhyay, K.K. A scheme of simultaneous cationic–anionic substitution in $CuCrO_2$ for transparent and superior p-type transport. *J. Phys. D Appl. Phys.* **2016**, *49*, 275109. [CrossRef]
138. Monteiro, J.F.H.L.; Monteiro, F.C.; Jurelo, A.R.; Mosca, D.H. Conductivity in (Ag,Mg)-doped delafossite oxide $CuCrO_2$. *Ceram. Int.* **2018**, *44*, 14101–14107. [CrossRef]
139. Wang, J.; Daunis, T.B.; Cheng, L.; Zhang, B.; Kim, J.; Hsu, J.W. Combustion synthesis of p-type transparent conducting $CuCro_{2+x}$ and Cu: Cro_x thin films at 180 °C. *ACS Appl. Mater. Interfaces* **2018**, *10*, 3732–3738. [CrossRef]
140. Jun, T.; Kim, J.; Sasase, M.; Hosono, H. Material design of p-type transparent amorphous semiconductor, Cu–Sn–I. *Adv. Mater.* **2018**, *30*, 1706573. [CrossRef] [PubMed]
141. Li, X.R.; Han, M.J.; Wu, J.D.; Shan, C.; Hu, Z.G.; Zhu, Z.Q.; Chu, J.H. Low voltage tunneling magnetoresistance in $CuCrO_2$-based semiconductor heterojunctions at room temperature. *J. Appl. Phys.* **2014**, *116*, 223701. [CrossRef]
142. Wang, J.; Lee, Y.-J.; Hsu, J.W. Sub-10 nm copper chromium oxide nanocrystals as a solution processed p-type hole transport layer for organic photovoltaics. *J. Mater. Chem. C* **2016**, *4*, 3607–3613. [CrossRef]
143. Ngo, T.; Palstra, T.; Blake, G. Crystallite size dependence of thermoelectric performance of $CuCrO_2$. *RSC Adv.* **2016**, *6*, 91171–91178. [CrossRef]
144. Ono, Y.; Satoh, K.-I.; Nozaki, T.; Kajitani, T. Structural, magnetic and thermoelectric properties of delafossite-type oxide, $CuCr_{1-x}Mg_xO_2$ ($0 \leq x \leq 0.05$). *Jpn. J. Appl. Phys.* **2007**, *46*, 1071. [CrossRef]
145. Li, X.; Han, M.; Zhang, X.; Shan, C.; Hu, Z.; Zhu, Z.; Chu, J. Temperature-dependent band gap, interband transitions, and exciton formation in transparent p-type delafossite $CuCr_{1-x}Mg_xO_2$ films. *Phys. Rev. B* **2014**, *90*, 035308. [CrossRef]
146. Yokobori, T.; Okawa, M.; Konishi, K.; Takei, R.; Katayama, K.; Oozono, S.; Shinmura, T.; Okuda, T.; Wadati, H.; Sakai, E. Electronic structure of the hole-doped delafossite oxides $CuCr_{1-x}Mg_xO_2$. *Phys. Rev. B* **2013**, *87*, 195124. [CrossRef]

147. Li, D.; Fang, X.; Deng, Z.; Zhou, S.; Tao, R.; Dong, W.; Wang, T.; Zhao, Y.; Meng, G.; Zhu, X. Electrical, optical and structural properties of $CuCrO_2$ films prepared by pulsed laser deposition. *J. Phys. D Appl. Phys.* **2007**, *40*, 4910. [CrossRef]
148. Benko, F.; Koffyberg, F. Preparation and opto-electronic properties of semiconducting $CuCrO_2$. *Mater. Res. Bull.* **1986**, *21*, 753–757. [CrossRef]
149. Nie, X.; Wei, S.-H.; Zhang, S. Bipolar doping and band-gap anomalies in delafossite transparent conductive oxides. *Phys. Rev. Lett.* **2002**, *88*, 066405. [CrossRef]
150. Santra, S.; Das, N.; Chattopadhyay, K. Sol–gel synthesis and characterization of wide band gap p-type nanocrystalline $CuBO_2$. *Mater. Lett.* **2013**, *92*, 198–201. [CrossRef]
151. Santra, S.; Das, N.; Chattopadhyay, K. Wide band gap p-type nanocrystalline $CuBO_2$ as a novel UV photocatalyst. *Mater. Res. Bull.* **2013**, *48*, 2669–2677. [CrossRef]
152. Santra, S.; Das, N.S.; Chattopadhyay, K.K. $CuBO_2$: A new photoconducting material. *AIP Conf. Proc.* **2013**, *1536*, 723–724. [CrossRef]
153. Renaud, A.; Chavillon, B.; Le Pleux, L.; Pellegrin, Y.; Blart, E.; Boujtita, M.; Pauporte, T.; Cario, L.; Jobic, S.; Odobel, F. $CuGaO_2$: A promising alternative for NiO in p-type dye solar cells. *J. Mater. Chem.* **2012**, *22*, 14353–14356. [CrossRef]
154. Xu, Z.; Xiong, D.; Wang, H.; Zhang, W.; Zeng, X.; Ming, L.; Chen, W.; Xu, X.; Cui, J.; Wang, M. Remarkable photocurrent of p-type dye-sensitized solar cell achieved by size controlled $CuGaO_2$ nanoplates. *J. Mater. Chem. A* **2014**, *2*, 2968–2976. [CrossRef]
155. Yu, M.; Natu, G.; Ji, Z.; Wu, Y. p-type dye-sensitized solar cells based on delafossite $CuGaO_2$ nanoplates with saturation photovoltages exceeding 460 mV. *J. Phys. Chem. Lett.* **2012**, *3*, 1074–1078. [CrossRef] [PubMed]
156. Renaud, A.l.; Cario, L.; Deniard, P.; Gautron, E.; Rocquefelte, X.; Pellegrin, Y.; Blart, E.; Odobel, F.; Jobic, S. Impact of Mg doping on performances of $CuGaO_2$ based p-type dye-sensitized solar cells. *J. Phys. Chem. C* **2013**, *118*, 54–59. [CrossRef]
157. Zhang, S.; Wei, S.-H.; Zunger, A. A phenomenological model for systematization and prediction of doping limits in II–VI and I–III–VI$_2$ compounds. *J. Appl. Phys.* **1998**, *83*, 3192–3196. [CrossRef]
158. Zhang, S.; Wei, S.-H.; Zunger, A. Microscopic origin of the phenomenological equilibrium "doping limit rule" in n-type III-V semiconductors. *Phys. Rev. Lett.* **2000**, *84*, 1232. [CrossRef]
159. Zhang, S.; Wei, S.-H.; Zunger, A. Intrinsic n-type versus p-type doping asymmetry and the defect physics of ZnO. *Phys. Rev. B* **2001**, *63*, 075205. [CrossRef]
160. Arai, T.; Iimura, S.; Kim, J.; Toda, Y.; Ueda, S.; Hosono, H. Chemical design and example of transparent bipolar semiconductors. *J. Am. Chem. Soc.* **2017**, *139*, 17175–17180. [CrossRef]
161. Cava, R.; Zandbergen, H.; Ramirez, A.; Takagi, H.; Chen, C.; Krajewski, J.; Peck, W., Jr.; Waszczak, J.; Meigs, G.; Roth, R. $LaCuO_{2.5+x}$ and $YCuO_{2.5+x}$ delafossites: Materials with triangular $Cu^{2+\delta}$ planes. *J. Solid State Chem.* **1993**, *104*, 437–452. [CrossRef]
162. Ingram, B.J.; Harder, B.J.; Hrabe, N.W.; Mason, T.O.; Poeppelmeier, K.R. Transport and defect mechanisms in cuprous delafossites. 2. $CuScO_2$ and $CuYO_2$. *Chem. Mater.* **2004**, *16*, 5623–5629. [CrossRef]
163. Ohta, H.; Orita, M.; Hirano, M.; Yagi, I.; Ueda, K.; Hosono, H. Electronic structure and optical properties of $SrCu_2O_2$. *J. Appl. Phys.* **2002**, *91*, 3074–3078. [CrossRef]
164. Kudo, A.; Yanagi, H.; Ueda, K.; Hosono, H.; Kawazoe, H.; Yano, Y. Fabrication of transparent p–n heterojunction thin film diodes based entirely on oxide semiconductors. *Appl. Phys. Lett.* **1999**, *75*, 2851–2853. [CrossRef]
165. Ohta, H.; Kawamura, K.; Orita, M.; Sarukura, N.; Hirano, M.; Hosono, H. UV-emitting diode composed of transparent oxide semiconductors: p-$SrCu_2O_2$/n-ZnO. *Electron. Lett* **2000**, *36*, 984–985. [CrossRef]
166. Hosono, H.; Ohta, H.; Hayashi, K.; Orita, M.; Hirano, M. Near-UV emitting diodes based on a transparent p–n junction composed of heteroepitaxially grown p-$SrCu_2O_2$ and n-ZnO. *J. Cryst. Growth* **2002**, *237*, 496–502. [CrossRef]
167. Inoue, S.-I.; Ueda, K.; Hosono, H.; Hamada, N. Electronic structure of the transparent p-type semiconductor (LaO)CuS. *Phys. Rev. B* **2001**, *64*, 245211. [CrossRef]
168. Ueda, K.; Inoue, S.; Hirose, S.; Kawazoe, H.; Hosono, H. Transparent p-type semiconductor: Lacuos layered oxysulfide. *Appl. Phys. Lett.* **2000**, *77*, 2701–2703. [CrossRef]
169. Hiramatsu, H.; Ueda, K.; Ohta, H.; Orita, M.; Hirano, M.; Hosono, H. Preparation of transparent p-type ($La_{1-x}Sr_xO$) cus thin films by RF sputtering technique. *Thin Solid Films* **2002**, *411*, 125–128. [CrossRef]

170. Hiramatsu, H.; Ueda, K.; Takafuji, K.; Ohta, H.; Hirano, M.; Kamiyama, T.; Hosono, H. Intrinsic excitonic photoluminescence and band-gap engineering of wide-gap p-type oxychalcogenide epitaxial films of LnCuOCh (Ln = La, Pr, and Nd; Ch = S or Se) semiconductor alloys. *J. Appl. Phys.* **2003**, *94*, 5805–5808. [CrossRef]
171. Ueda, K.; Hosono, H.; Hamada, N. Energy band structure of LnCuOCh (Ch = S, Se and Te) calculated by the full-potential linearized augmented plane-wave method. *J. Phys. Condens. Matter* **2004**, *16*, 5179. [CrossRef]
172. Liu, M.-L.; Wu, L.-B.; Huang, F.-Q.; Chen, L.-D.; Chen, I.-W. A promising p-type transparent conducting material: Layered oxysulfide [Cu_2S_2][$Sr_3Sc_2O_5$]. *AIP J. Appl. Phys.* **2007**, *102*, 116108. [CrossRef]
173. Scanlon, D.O.; Buckeridge, J.; Catlow, C.R.A.; Watson, G.W. Understanding doping anomalies in degenerate p-type semiconductor lacuose. *J. Mater. Chem. C* **2014**, *2*, 3429–3438. [CrossRef]
174. Ueda, K.; Takafuji, K.; Hiramatsu, H.; Ohta, H.; Kamiya, T.; Hirano, M.; Hosono, H. Electrical and optical properties and electronic structures of LnCuOS (Ln = La∼Nd). *Chem. Mater.* **2003**, *15*, 3692–3695. [CrossRef]
175. Hiramatsu, H.; Ueda, K.; Ohta, H.; Hirano, M.; Kamiya, T.; Hosono, H. Degenerate p-type conductivity in wide-gap $LaCuOS_{1-x}Se_x$ (x = 0–1) epitaxial films. *Appl. Phys. Lett.* **2003**, *82*, 1048–1050. [CrossRef]
176. Hiramatsu, H.; Yanagi, H.; Kamiya, T.; Ueda, K.; Hirano, M.; Hosono, H. Crystal structures, optoelectronic properties, and electronic structures of layered oxychalcogenides MCuOCh (M = Bi, La; Ch = S, Se, Te): Effects of electronic configurations of M^{3+} ions. *Chem. Mater.* **2007**, *20*, 326–334. [CrossRef]
177. Ueda, K.; Hosono, H.; Hamada, N. Valence-band structures of layered oxychalcogenides, LaCuOCh (Ch = S, Se, and Te), studied by ultraviolet photoemission spectroscopy and energy-band calculations. *J. Appl. Phys.* **2005**, *98*, 043506. [CrossRef]
178. Ueda, K.; Takafuji, K.; Yanagi, H.; Kamiya, T.; Hosono, H.; Hiramatsu, H.; Hirano, M.; Hamada, N. Optoelectronic properties and electronic structure of YCuOSe. *J. Appl. Phys.* **2007**, *102*, 113714. [CrossRef]
179. Hiramatsu, H.; Orita, M.; Hirano, M.; Ueda, K.; Hosono, H. Electrical conductivity control in transparent p-type (LaO)CuS thin films prepared by rf sputtering. *J. Appl. Phys.* **2002**, *91*, 9177–9181. [CrossRef]
180. Ueda, K.; Hosono, H. Band gap engineering, band edge emission, and p-type conductivity in wide-gap $LaCuOS_{1-x}Se_x$ oxychalcogenides. *J. Appl. Phys.* **2002**, *91*, 4768–4770. [CrossRef]
181. Hiramatsu, H.; Ueda, K.; Ohta, H.; Orita, M.; Hirano, M.; Hosono, H. Heteroepitaxial growth of a wide-gap p-type semiconductor, LaCuOS. *Appl. Phys. Lett.* **2002**, *81*, 598–600. [CrossRef]
182. Hamberg, I.; Granqvist, C.G. Evaporated Sn-doped In_2O_3 films: Basic optical properties and applications to energy-efficient windows. *J. Appl. Phys.* **1986**, *60*, R123–R160. [CrossRef]
183. Slaoui, A.; Siffert, P. Determination of the electron effective mass and relaxation time in heavily doped silicon. *Phys. Status Solidi (a)* **1985**, *89*, 617–622. [CrossRef]
184. Hosono, H. Built-in nanostructures in transparent oxides for novel photonic and electronic functions materials. *Int. J. Appl. Ceram. Technol.* **2004**, *1*, 106–118. [CrossRef]
185. Moss, T. The interpretation of the properties of indium antimonide. *Proc. Phys. Soc. Sect. B* **1954**, *67*, 775. [CrossRef]
186. Goto, Y.; Tanaki, M.; Okusa, Y.; Shibuya, T.; Yasuoka, K.; Matoba, M.; Kamihara, Y. Effects of the Cu off-stoichiometry on transport properties of wide gap p-type semiconductor, layered oxysulfide lacuso. *Appl. Phys. Lett.* **2014**, *105*, 022104. [CrossRef]
187. Hiramatsu, H.; Ueda, K.; Ohta, H.; Kamiya, T.; Hirano, M.; Hosono, H. Excitonic blue luminescence from p-LaCuOSe/n-$InGaZn_5O_8$ light-emitting diode at room temperature. *Appl. Phys. Lett.* **2005**, *87*, 211107. [CrossRef]
188. Yanagi, H.; Kikuchi, M.; Kim, K.-B.; Hiramatsu, H.; Kamiya, T.; Hirano, M.; Hosono, H. Low and small resistance hole-injection barrier for NPB realized by wide-gap p-type degenerate semiconductor, LaCuOSe:Mg. *Org. Electron.* **2008**, *9*, 890–894. [CrossRef]
189. Zhang, N.; Liu, X.; Shi, D.; Tang, B.; Annadi, A.; Gong, H. Achievement of highly conductive p-type transparent ndcuos film with cu deficiency and effective doping. *Mater. Today Chem.* **2018**, *10*, 79–89. [CrossRef]
190. Kremers, H.; Stevens, R. Observations on the rare earths. XIV. The preparation and properties of metallic lanthanum. *J. Am. Chem. Soc.* **1923**, *45*, 614–617. [CrossRef]
191. Nakachi, Y.; Ueda, K. Single crystal growth of lacuos by the flux method. *J. Cryst. Growth* **2008**, *311*, 114–117. [CrossRef]

192. Doussier-Brochard, C.; Chavillon, B.; Cario, L.; Jobic, S. Synthesis of p-type transparent LaOCuS nanoparticles via soft chemistry. *Inorg. Chem.* **2010**, *49*, 3074–3076. [CrossRef] [PubMed]
193. Lian, J.; Li, N.; Wang, H.; Su, Y.; Zhang, G.; Liu, F. Synthesis of lacuos nanopowder by a novel precipitation combined with reduction route. *Ceram. Int.* **2016**, *42*, 11473–11477. [CrossRef]
194. Zhang, N.; Gong, H. p-type transparent lacuos semiconductor synthesized via a novel two-step solid state reaction and sulfurization process. *Ceram. Int.* **2017**, *43*, 6295–6302. [CrossRef]
195. Hiramatsu, H.; Ueda, K.; Ohta, H.; Hirano, M.; Kamiya, T.; Hosono, H. Wide gap p-type degenerate semiconductor: Mg-doped LaCuOSe. *Thin Solid Films* **2003**, *445*, 304–308. [CrossRef]
196. Zhang, N.; Shi, D.; Liu, X.; Annadi, A.; Tang, B.; Huang, T.J.; Gong, H. High performance p-type transparent LaCuOS thin film fabricated through a hydrogen-free method. *Appl. Mater. Today* **2018**, *13*, 15–23. [CrossRef]
197. Wang, Z.; Nayak, P.K.; Caraveo-Frescas, J.A.; Alshareef, H.N. Recent developments in p-type oxide semiconductor materials and devices. *Adv. Mater.* **2016**, *28*, 3831–3892. [CrossRef] [PubMed]
198. Xu, W.; Li, H.; Xu, J.-B.; Wang, L. Recent advances of solution-processed metal oxide thin-film transistors. *ACS Appl. Mater. Interfaces* **2018**, *10*, 25878–25901. [CrossRef]
199. Nolan, M.; Elliott, S.D. The p-type conduction mechanism in Cu_2O: A first principles study. *Phys. Chem. Chem. Phys.* **2006**, *8*, 5350–5358. [CrossRef]
200. Fortunato, E.; Figueiredo, V.; Barquinha, P.; Elamurugu, E.; Barros, R.; Gonçalves, G.; Park, S.-H.K.; Hwang, C.-S.; Martins, R. Thin-film transistors based on p-type Cu_2O thin films produced at room temperature. *Appl. Phys. Lett.* **2010**, *96*, 192102. [CrossRef]
201. Saji, K.J.; Populoh, S.; Tiwari, A.N.; Romanyuk, Y.E. Design of p-CuO/n-ZnO heterojunctions by rf magnetron sputtering. *Phys. Status Solidi (a)* **2013**, *210*, 1386–1391. [CrossRef]
202. Chen, W.-C.; Hsu, P.-C.; Chien, C.-W.; Chang, K.-M.; Hsu, C.-J.; Chang, C.-H.; Lee, W.-K.; Chou, W.-F.; Hsieh, H.-H.; Wu, C.-C. Room-temperature-processed flexible n-InGaZnO/p-Cu_2O heterojunction diodes and high-frequency diode rectifiers. *J. Phys. D Appl. Phys.* **2014**, *47*, 365101. [CrossRef]
203. Dhakal, D.; Waechtler, T.; Schulz, S.E.; Gessner, T.; Lang, H.; Mothes, R.; Tuchscherer, A. Surface chemistry of a Cu(I) beta-diketonate precursor and the atomic layer deposition of Cu_2O on SiO_2 studied by X-ray photoelectron spectroscopy. *J. Vac. Sci. Technol. A* **2014**, *32*, 041505. [CrossRef]
204. Kwon, J.-D.; Kwon, S.-H.; Jung, T.-H.; Nam, K.-S.; Chung, K.-B.; Kim, D.-H.; Park, J.-S. Controlled growth and properties of p-type cuprous oxide films by plasma-enhanced atomic layer deposition at low temperature. *Appl. Surf. Sci.* **2013**, *285*, 373–379. [CrossRef]
205. Lee, S.W.; Lee, Y.S.; Heo, J.; Siah, S.C.; Chua, D.; Brandt, R.E.; Kim, S.B.; Mailoa, J.P.; Buonassisi, T.; Gordon, R.G. Improved Cu_2O-based solar cells using atomic layer deposition to control the Cu oxidation state at the p-n junction. *Adv. Energy Mater.* **2014**, *4*, 1301916. [CrossRef]
206. Chen, A.; Long, H.; Li, X.; Li, Y.; Yang, G.; Lu, P. Controlled growth and characteristics of single-phase Cu_2O and Cuo films by pulsed laser deposition. *Vacuum* **2009**, *83*, 927–930. [CrossRef]
207. Tanaka, H.; Shimakawa, T.; Miyata, T.; Sato, H.; Minami, T. Electrical and optical properties of TCO–Cu_2O heterojunction devices. *Thin Solid Films* **2004**, *469*, 80–85. [CrossRef]
208. Kikuchi, N.; Tonooka, K. Electrical and structural properties of Ni-doped Cu_2O films prepared by pulsed laser deposition. *Thin Solid Films* **2005**, *486*, 33–37. [CrossRef]
209. Zou, X.; Fang, G.; Yuan, L.; Li, M.; Guan, W.; Zhao, X. Top-gate low-threshold voltage p-Cu_2O thin-film transistor grown on SiO_2/Si substrate using a high-κ HfON gate dielectric. *IEEE Electron Device Lett.* **2010**, *31*, 827–829. [CrossRef]
210. Jeong, S.; Aydil, E.S. Structural and electrical properties of Cu_2O thin films deposited on ZnO by metal organic chemical vapor deposition. *J. Vac. Sci. Technol. A* **2010**, *28*, 1338–1343. [CrossRef]
211. Park, I.-J.; Jeong, C.-Y.; Myeonghun, U.; Song, S.-H.; Cho, I.-T.; Lee, J.-H.; Cho, E.-S.; Kwon, H.-I. Bias-stress-induced instabilities in p-type Cu_2O thin-film transistors. *IEEE Electron Device Lett.* **2013**, *34*, 647–649. [CrossRef]
212. Nie, S.; Sun, J.; Gong, H.; Chen, Z.; Huang, Y.; Xu, J.; Zhao, L.; Zhou, W.; Wang, Q. Glucose-assisted reduction achieved transparent p-type cuprous oxide thin film by a solution method. *EPL (Eur. Lett.)* **2016**, *115*, 37005. [CrossRef]
213. Kim, S.Y.; Ahn, C.H.; Lee, J.H.; Kwon, Y.H.; Hwang, S.; Lee, J.Y.; Cho, H.K. p-Channel oxide thin film transistors using solution-processed copper oxide. *ACS Appl. Mater. Interfaces* **2013**, *5*, 2417–2421. [CrossRef] [PubMed]

214. Yu, W.; Han, M.; Jiang, K.; Duan, Z.; Li, Y.; Hu, Z.; Chu, J. Enhanced fröhlich interaction of semiconductor cuprous oxide films determined by temperature-dependent raman scattering and spectral transmittance. *J. Raman Spectrosc.* **2013**, *44*, 142–146. [CrossRef]
215. Matsuzaki, K.; Nomura, K.; Yanagi, H.; Kamiya, T.; Hirano, M.; Hosono, H. Epitaxial growth of high mobility Cu_2O thin films and application to p-channel thin film transistor. *Appl. Phys. Lett.* **2008**, *93*, 202107. [CrossRef]
216. Yao, Z.; Liu, S.; Zhang, L.; He, B.; Kumar, A.; Jiang, X.; Zhang, W.; Shao, G. Room temperature fabrication of p-channel Cu_2O thin-film transistors on flexible polyethylene terephthalate substrates. *Appl. Phys. Lett.* **2012**, *101*, 042114. [CrossRef]
217. Pattanasattayavong, P.; Thomas, S.; Adamopoulos, G.; McLachlan, M.A.; Anthopoulos, T.D. p-Channel thin-film transistors based on spray-coated Cu_2O films. *Appl. Phys. Lett.* **2013**, *102*, 163505. [CrossRef]
218. Liu, A.; Liu, G.; Zhu, H.; Song, H.; Shin, B.; Fortunato, E.; Martins, R.; Shan, F. Water-induced scandium oxide dielectric for low-operating voltage n-and p-type metal-oxide thin-film transistors. *Adv. Funct. Mater.* **2015**, *25*, 7180–7188. [CrossRef]
219. Liu, A.; Nie, S.; Liu, G.; Zhu, H.; Zhu, C.; Shin, B.; Fortunato, E.; Martins, R.; Shan, F. In situ one-step synthesis of p-type copper oxide for low-temperature, solution-processed thin-film transistors. *J. Mater. Chem. C* **2017**, *5*, 2524–2530. [CrossRef]

© 2019 by the authors. Licensee MDPI, Basel, Switzerland. This article is an open access article distributed under the terms and conditions of the Creative Commons Attribution (CC BY) license (http://creativecommons.org/licenses/by/4.0/).

Review

Coatings in Photovoltaic Solar Energy Worldwide Research

Nuria Novas, Alfredo Alcayde, Dalia El Khaled and Francisco Manzano-Agugliaro *

Department of Engineering, ceiA3, University of Almeria, 04120 Almeria, Spain; nnovas@ual.es (N.N.); aalcayde@ual.es (A.A.); dalia.elkhaled@gmail.com (D.E.K.)
* Correspondence: fmanzano@ual.es

Received: 23 October 2019; Accepted: 23 November 2019; Published: 27 November 2019

Abstract: This paper describes the characteristics of contributions that were made by researchers worldwide in the field of Solar Coating in the period 1957–2019. Scopus is used as a database and the results are processed while using bibliometric and analytical techniques. All of the documents registered in Scopus, a total of 6440 documents, have been analyzed and distributed according to thematic subcategories. Publications are analyzed from the type of publication, field of use, language, subcategory, type of newspaper, and the frequency of the keyword perspectives. English (96.8%) is the language that is most used for publications, followed by Chinese (2.6%), and the rest of the languages have a less than < 1% representation. Publications are studied by authors, affiliations, countries of origin of the authors, and H-index, which it stands out that the authors of China contribute with 3345 researchers, closely followed by the United States with 2634 and Germany with 1156. The Asian continent contributes the most, with 65% of the top 20 affiliations, and Taiwan having the most authors publishing in this subject, closely followed by Switzerland. It can be stated that research in this area is still evolving with a great international scientific contribution in improving the efficiency of solar cells.

Keywords: solar energy; coatings; scopus; material solar cell; thin film; polycrystalline; organic solar cell; thin film a-Si: H; optical design; light trapping

---

1. Introduction

Energy needs are a global growing problem in the era of technology. Citizens and governments are gradually becoming aware of the sustainable use of world resources [1]. Many are the developments in energy systems based on renewable energy, such as wind, photovoltaic, biomass, nuclear, etc., implemented on both a small and high scale. Renewable energy resources largely depend on the climate of the site; different renewable energies could be applied in different regions. Society demands clean and sustainable energy; this implies research in efficient clean energy. Among existing different renewable energies, solar energy is one of the most attractive for future energy sources [2–4] and photovoltaics is the most implemented one. Photovoltaic applications are very diverse, and they range from the incorporation into consumer products, such as watches, calculators, battery chargers, and a multitude of products from the leisure industry. They can also be applied in small-scale systems, like remote installations in structures, called solar gardens, or systems applied to the industrial and domestic facilities for small villages and water pumping stations. Not forgetting the large power production stations for supply of network connection. Energy policies play an important role in the development of renewable energy [5,6].

Currently, with the arrival of intelligent and sustainable buildings, solar modules that are installed in the building are installed in both roofs and part of the facade and windows, where, apart from energy efficiency, the aesthetics of the architecture are considered. Transparent and biphasic thin film solar modules contribute to their application in these structures [7].

These photovoltaic systems depend, to a large extent on the physical and chemical properties of their materials, the wavelength of the captured light, its intensity, and its angle of incidence, the characteristics of the surface or texture as well as the presence or absence of superficial coatings. In addition to these factors, temperature, pressure, ease of processing, durability, price, and costs throughout the life are important in material selection. Photovoltaic energy has been highly researched in the last 60 years, with the intention of reducing manufacturing costs and, at the same time, improving performance. The improvement of maintenance (protection against abrasion, corrosion, cleaning, etc.), increase in the life of the components, and incorporation of materials based on plastics and underlying substrates as coatings are among the cost reduction factors.

The starting silicon wafer is one of the main costs of silicon photovoltaic cells; the degree of purity largely defines the performance of the cell. This has led to the solar cells with nanostructure p-n radial junctions, where the quantity of Si and its quality is reduced. Improved light absorption in ultrafine solar silicon film is important in improving efficiency and reducing costs [8–10]. Thin-layer technologies also use less Si, reducing the production costs, although with limited efficiency, which increases the total system costs.

Research is being conducted for improving the capture of light in order to reduce the thickness of the layer, which entails reducing the material, and improving the efficiency, which has an impact on manufacturing costs. In this sense, solar cells have been improved by advances in diffractive optical elements (DOE) that are used in many areas of optics, such as spectroscopy and interferometry, among others. The shape of the grid slot can be used as an optimization parameter for specific tasks. In many cases, DOEs are manufactured on flat substrates for simplicity, but they offer many important additional advantages on curved surfaces [11]. With the development of computers and their application to different fields, such as homography, techniques such as interferometric recording have been developed. Digital homography (computer-generated holograms) has allowed for great flexibility in creating forms in substrates with high precision [12].

For this, nanostructures have been designed in different ways, depending on the type of solar cells. The compromise between optical and electrical performance currently limits solar cells. There are different proposals regarding whether nanostructures should be periodic or random. Non-fullerene acceptors (NFA) become an interesting family of organic photovoltaic materials and they have attracted considerable interest in their great potential in manufacturing large surface flexible solar panels through low-cost coating methods [13].

Research regarding the improvements in Solar Coating are in continuous evolution with the incorporation of new materials, structures, and the growing demand for energy; all these advances are mainly focused on improving the efficiency of photovoltaic panels. From this point of view, there are several scientific communities making continuous contributions from different fields. These contributions are doubled per decade, which entails a huge number of documents to deal with. The documents within the same field of research are distributed in scientific communities that are promoted through the interrelations between the authors and their publications. The collaboration of the authors in different communities makes the progress of science more productive, since there are not only research relationships between authors, but also between institutions that support the necessary tests with their laboratories and facilities. This exponentially increases the progress in science and technology. In this work, we study the different communities that have consolidated over time and the relationships between them.

2. Materials and Methods

This paper analyzes all of the scientific publications indexed on Scopus data base that deal with Solar Coating. There are search engines on the web based on Scientometric indicators, such as number and quality of contributions, according to the metric of the journal or the author. The results of these searches do not measure the relationships between the authors; this limits the establishment of collaborative communities. Technology, like science, advances through continuous collaborations between public or private research entities; therefore, it is important to develop metrics that incorporate the authors' relationships. There are different studies that carry out comparisons between Scopus and Web of Science, and they reach the conclusion that Scopus is the scientific database with the greatest contributions [14,15]. In addition, Scopus allows for the development of APIs (Application Programming Interface) that directly extract information from the database, allowing for an analysis of them [16]. Figure 1 shows the API developed, as it can be considered as the core of the methodology of this manuscript. Accordingly, a search for keywords related to Solar Coating has been carried out to find global relations between the generated communities, their authors, and research institutions. The search is performed for TITLE-ABS-KEY ("advanced glazing*" OR "Solar window*" OR "light trapping" OR "diffractive element*") obtaining many documents and their relationships. This requires a debugging process to avoid unnecessary information that prevents an overview, which reduces the number of documents and their relationships. Documents that have no relations within the generated communities are eliminated in the debugging process. The final data set was analyzed while using statistical tools that were based on diagrams and presentation of the data processed. The open source tool, like Gephi (https://gephi.org), was used, which incorporates statistical resources and data visualization, mainly the algorithm ForceAtlas2 [17]. In this way, the different clusters were automatically identified. After this, the information of each cluster was analyzed in Excel, while using the dynamic data tables and the word cloud has been realized with the software Word Art (https://wordart.com/create). Note that the size of the keyword must be proportional to its frequency and the number of times that keyword appears in the analyzed articles in a representation by cloud of words.

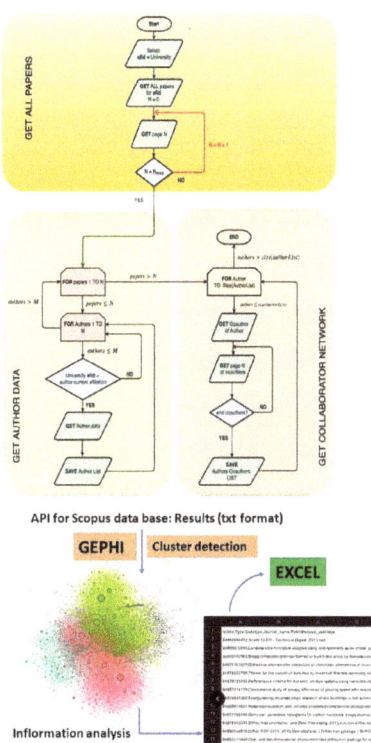

Figure 1. Flow diagram of the API that allowed for extracting the information of Scopus database.

3. Results

3.1. Communities Detection

A total of 6440 documents are obtained with a total of 21,301 relations between the authors after searching for the keywords. After the debugging process to avoid unnecessary information, documents are reduced by 39.1% and relations by 2.12%. Figure 2 shows the 3924 documents with 20,849 relationships that were obtained after the process of purification and statistical treatment. Figure 2 shows the distribution of the six detected communities that publish in Solar Coating topics with the Gephi program. As you can see, there is a main nucleus that is formed by five communities and another exterior formed by a single community. In Figure 2a, a node represents each publication and the size of the node is a function of their relationships, so that it shows the frequency with which the node appears in the shortest path between two randomly selected nodes in/between communities, showing the influence of the author within the community. In this way, not only the common metrics in search engines, such as Google Scholar, are considered, but also the collaborations between the authors. The size of a node varies according to its relationships to indicate the most influential nodes. The reason why an author who has a highly referenced and published document, but who works by himself, will only have a smaller node than a less referenced author with greater collaborations.

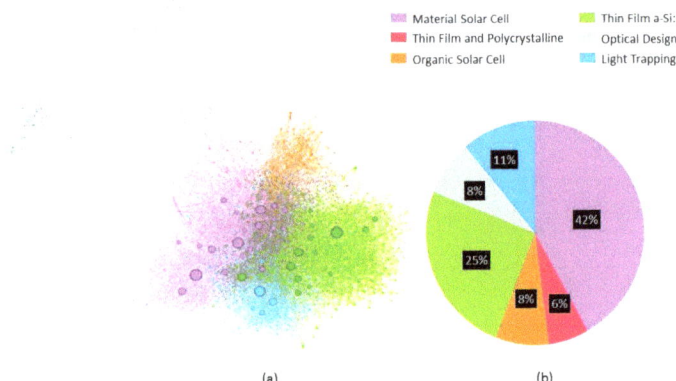

Figure 2. Representation of the communities investigating about "solar coating": (**a**) represent the interaction of the communities as a whole; and, (**b**) Representation for the distribution of the percentage of the communities.

Figure 2b presents the contribution in percentage of each community, since it is difficult to see the total size of each community due to the interrelation in Figure 2a. There are two communities that stand out for their size and they are the Material Solar Cell and Thin Fill Cells a-Si: H community. Community 0 (Material Solar Cell) is the largest with 42.2% of total publications. In this community, you can see the highest concentration of related nodes, where it publishes the advances on the materials used to improve the capture of light. The Thin Fill Cells a-Si: H community publishes 25.36% on the improvements in amorphous cells of hydrolyzed silicon. This community, besides being the second largest, is also the one that has a large concentration of authors that are related to other nodes, as it can be seen in Figure 2.

Figure 3 shows a cloud words of the global keywords obtained in the search. The most used keyword is "photovoltaic cells", with 147 times within the Material Solar Cell community, followed by "Thin film solar cells" with 96 times from the Thin film a-Si: H community. The third most used is "Silicon", also from the Solar Cell Material community with 85 times. The three words belong to the two communities with the largest number of publications, as shown in Figure 2b. Globally, the most repeated words are the most generic, such as "Film", "Thin", and "ZnO", as shown in Figure 3.

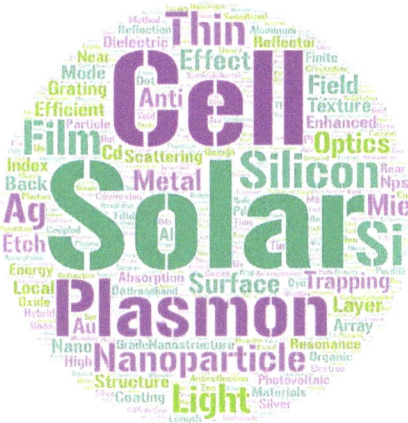

Figure 3. Cloud words of the keywords got in the global search.

3.2. Analysis of the Communities

Each community maintains a common theme, although being very interrelated with the rest of the communities. Next, the main nodes of each community and their most significant contributions are analyzed; this will allow for us to understand their research theme.

Community 0 (Material Solar Cell) investigates how to improve the capture of light by using different structures and materials used. Figure 4a shows the most representative keywords of the Solar Cell Material Community, which shows the number of times and their percentage of repetition within the community. The most representative word is "Photovoltaic cells", followed by "Silicon"; these words are very generic in the subject of photovoltaic solar energy and hints that this community is the widest when publishing more generic developments from which other more specific communities are nurtured. Hence, it has three important nodes that are very referenced, as shown in Figure 4b, and a multitude of publications that are very referenced not only by this community, but by the remaining ones, as it can be seen in Figure 2. Publications of the most referenced nodes in Scopus in order of size are:

- "Light trapping in silicon nanowire solar cells" [18] with a total of 1572 cites.
- "Fundamental limit of nanophotonic light trapping in solar cells" [19] with 586 cites.
- "Improving thin-film crystalline silicon solar cell efficiencies with photonic crystals" [20] with 532 cites.

Si cable assemblies are studied in this community. These are an interesting architecture for solar energy collection applications, and they can offer a mechanically flexible alternative to Si wafers for photovoltaic energy. Cables must absorb sunlight in a wide range of wavelengths and angles of incidence to achieve competitive conversion efficiencies, despite only occupying a modest fraction of the array volume. These matrices show a better near-infrared absorption, which allows for its absorption of sunlight to exceed the ray optics that traps the light absorption limit for an equivalent volume of textured plane, over a wide range of angles of incidence. The geometry of the cable network, together with other nanostructured geometries, offers opportunities to manipulate the relationship between the lighting area and the volume of absorption being useful in improving the efficiency or reducing the consumption of materials of many photovoltaic technologies. Kelzenberg et al. [21] showed that arrays presenting less than 5% of the cable area fraction can reach up to 96% of maximum absorption, and that they can absorb up to 85% of the substances integrated in the day, above the direct sunlight band. Garnett et al. [18] developed a structure of nanowires with large radial surface photovoltaic splicing p-n with efficiencies between 5% and 6%.

Material Solar Cell		
Keys	Nº	%
Photovoltaic cells	147	3.1
Silicon	85	1.8
Antireflection	29	0.6
Nanostructures	29	0.6
Photonic crystals	28	0.6
Silicon nanowires	26	0.6
Texturing	26	0.6
Photon management	25	0.5
Photonic cristal	23	0.5
Solar energy	23	0.5

(a) (b)

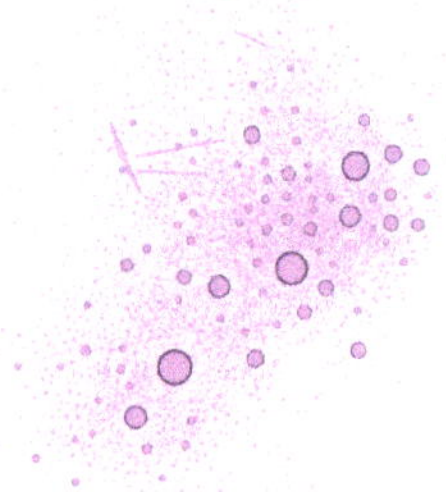

Figure 4. Representation of the Material Solar Cell community: (**a**) keywords; and, (**b**) isolated distribution of the publications.

Brongersma el al. [22] review the theory of nanophotonic light capture in periodic structures. Light collection schemes can be used to improve absorption in photovoltaic (PV) cells. They help to increase cell efficiency and reduce the production costs. In a homogeneous bulk cell with reflection mirror backing, (in a homogeneous bull cell with a back reflection mirror) the maximum enhancement factor attainable by the light trapping schemes is $4n^2/\sin^2(\theta)$, where n is the index of refraction of the material and θ is half of the apex angle of the absorption cone. Ultrafine cells with efficiencies that can exceed the traditional 4n2 limit are investigated. It involves the development of new computational tools that are capable of operating in the domain of wave optics, dealing with non-periodic structures and performing a joint electrical and optical optimization [23]. Yu et al. [19] studied the case of the capture of light in grid structures with periodicity at the wavelength scale. Light capture can improve cell efficiency, because thinner cells provide a better collection of photogenerated cells and potentially higher open circuit voltage. Yu et al. [19,24] developed "a statistical coupled-mode theory for nanophotonic light trapping" theory. Yu et al. [24], this theory is applied to the one-dimensional (1D) and two-dimensional (2D) grids that have close or even smaller thicknesses than the wavelength of the light and conclude that the 2D grids have a greater improvement factor. Yang et al. [25] used the coupled wave analysis method for textured sub length wavelength (STDS), which are important in obtaining high efficiency, due to their almost perfect anti-reflective properties.

Another author's study method was based on geometric optics and wave optics applied to thin-film crystalline silicon solar cell [20]. They manage to increase efficiency with the use of photonic glass, increasing 24.0% in an optimized 1D to 31.3% by adding an optimized 2D grid.

Wang et al. [26] present a double-sided grid design, in which the front and rear surfaces of the cell are separately optimized for antireflection and light capture, respectively. The authors propose a structure based on nano cones of different sizes for the upper layer (the period is 500 nm, the base radius is 250 nm, and the height is 710 nm) and lower layer (the period is 1000 nm, the base radius is

475 nm, and the height is 330 nm). Their experimental results approximate the limit of the theoretical absorption spectrum of Yablonovitch.

Community 1 (Thin Film and Polycrystalline) publishes the advances in the efficiency of thin cells and thin polycrystalline cells. Figure 5a shows the most representative keywords of the Thin Film and Polycrystalline community, which shows the number of times and their percentage of repetition within the community. The most representative word is "Silicon", followed by "Crystalline silicon" and "Epitaxy"; these words are generic of all communities. This community has a lot of keywords; it is the smallest and therefore its most repeated keywords are the most generic, the rest are more focused on the specific theme of the community. Figure 5b displays the distribution of published documents. Unlike the Material Solar Cell community, this is much more specific and, although it maintains connections with other communities, its articles do not have references from the Material Solar Cell and Thin film a-Si community: H.

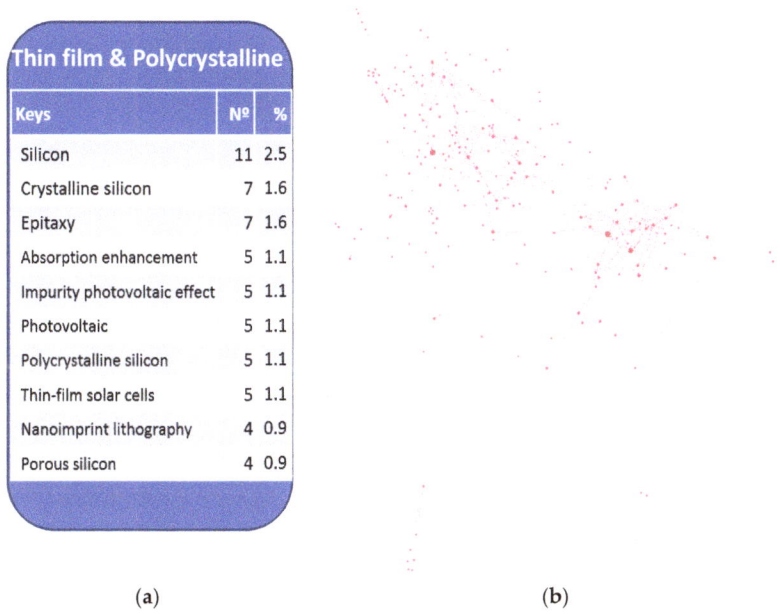

(a) (b)

Figure 5. Representation of Thin film and Polycrystalline community: (**a**) keywords; and, (**b**) isolated distribution of the publications.

In this community, a lot of research is being carried out to reduce the consumption of Si per watt peak. In addition to reducing the cost, a reduction in the thickness of the solar cell theoretically allows for an increase in the performance of the device. The long-term stability of thin film photovoltaic modules is increasing, while also reducing costs [27].

Thin film-based technologies show much lower surface production costs than bulk Si PV. Becker et al. [28] show the development of the i2 modules and the challenges that are faced by high-quality crystalline Si cells of thin film on glass, with the main ones being improvements in light trapping characteristics, low temperature junction processing, and cell metallization. Xue et al. [29] propose the Liquid Phase Crystallization Techniques (LPC) in the manufacture of high-quality crystalline silicon thin film solar cells in glass. Therefore, LPC is used for the development of double-sided silicon films, and different nanophotonic geometries of light capture are studied, concluding that this 10 mm thick double-sided silicon films can present maximum short-circuit current densities that are achievable in solar cells up to 38mA/cm^2 while assuming zero-parasite absorption.

Improvements in light entrapment in Si Polycristalline thin-layer solar cells (pc-Si) are based on the random scattering of light in the absorbent layer by glass substrate texture, or silicon film etching texture and plasmonic nanoparticles [29]. The thin-layer solar cells pc-Si on glass offer the possibility of achieving efficiencies of a single union of 15%. This is achieved by developing structures that improve light entrapment, being mainly based on silicon nanostructures, such as porous silicon, nanowires of silicon, and nano-silicon holes [29]. [30,31] proposes using a "seed layer" to obtain a high quality material, the use of ZnO and aluminum cladding as a method of improving light collection, and the use of high quality materials for the evaporation of the electron beam for the deposition of the absorbers, which offers a high potential for cost reduction, to obtain efficiency improvements and a reduction in costs. Another proposal is to use nanowire matrices to improve light entrapment and the design of the cell structure to minimize parasitic absorption, together with suppression of surface recombination, while using a multi-HIT configuration (hetero junction with intrinsic thin layer) core-based solar-based nanowire cells that were prepared in the thin film of low-cost pc-Si, developing an 8 μm pc-Si cell [32].

Another method that is based on surface plasmonic resonance (SPR) and a periodic hybrid matrix composed of a graphene ring at the top of the absorbent layer separated by an insulating layer to achieve an improvement of multiband absorption, increases the basis for simultaneous photodetection at multiple wavelengths with high efficiency and tunable spectral selectivity [33].

In [34], they propose a complete method for studying long-term light entrapment, the use of quantum efficiency data, and expressions of the calculation of Z0 and RBACK (reflectivity of the rear reflector defined in [10] for any solar cell), where Z0 is the optical path of short band length factor Z0 of Rand and Basore [35], and it is a multiple of the thickness of the cell necessary to generate equal to that found in the device. Although there are not very relevant nodes as compared to the others, it should be noted that the publications of the most cited nodes in Scopus for this scientific community in order of size are:

- "Polycrystalline silicon thin-film solar cells: Status and perspectives" [36] with a total of 117 cites.
- "Crystalline thin-foil silicon solar cells: Where crystalline quality meets thin-film processing" [37] with 64 cites.
- "Double-side textured liquid phase crystallized silicon thin-film solar cells on imprinted glasswith" [38] 37 times cited.

Community 2 (Organic solar cells) investigates an alternative to silicon-based photovoltaic cells, organic solar cells (OSC), or also called organic photovoltaic cells (OPV). Figure 6a shows the most representative keywords of the Organic solar cell's community, which shows the number of times and their percentage of repetition within the community. One of the most representative words is "Organic solar cells", after which the community is named. The following words are specific to the topic treated in this community, such as: "Organic photovoltaics", "Light harvesting", and "Polymer solar cell". Despite the small representativeness, 8.3% of the publications, (Figure 2), this community has a greater concentration of publications with references, as it can be seen in the size of the circles in Figure 6b, which is unlike the community Thin Film and Polycrystalline. This community has ties with the rest of the communities.

OSC cells have interesting advantages due to their characteristics, such as their lightweight, flexibility, and possibility of producing them profitably for large surfaces. These features have made of these cells very valid for applications in electronic textiles, synthetic leather, and robot, etc. The main disadvantage is their low energy conversion efficiency, which is mainly because the light absorption properties in an organic active layer have short optical absorption lengths ($L_A \sim 100$ nm) and exciton diffusion length ($L_D \sim 10$ nm). This implies that a reduction in the thickness of the active layer affects deterioration in performance, but an increase in thickness implies an increase in the series resistance and reduction in the collection of carriers. Therefore, a compromise between both of the situations is sought, efficient light collection and efficient load collection. The optical optimization that is used in other thin-layer technologies can be useful in achieving' maximum concentration in the absorbent layer, some of the proposals for improvement in light capture are based on modifying the structure, mainly plasmonic nanostructures, where photonic crystals are used, metal gratings, buried nanoelectrodes, etc.; however, in essence, they increase the organic surface layer [39,40]. Ko et al. [41] study the different nanostructure-based uptake systems for OSC cells while using both Plasmon surfaces and anti-reflective coatings, and photonic crystal (PC) nanostructure. Although theoretical calculations suggest that the efficiency increases the accumulation of absorption of light in nanostructured devices, the results show that they are still inferior to the highest reported conventional organic photovoltaic cells, which implies that further research in this field must be carried out to obtain thinner layers of photoactive material that improve the performance. The use of metallic nanomaterials can improve the capture of light in OSC and, although most nanoparticles (NPs) limit the improvement of the efficiency of power conversion to a narrow spectral range, broadband capture is desirable. The proposal of Li et al. [42] is the combination of Ag nanomaterials in different ways (Localized plasmonic resonances (LPRs), Ag nano prisms, and NPs mixed with Ag) for better broadband absorption and increased short-circuit photocurrent density. Out of the three experiments, the one with NPs mixed with Ag is the one with the highest efficiency with power conversion efficiency of 4.3%. They conclude that the cooperative plasmonic effects in metallic nanomaterials with different types of materials, shapes, size, and even the polarization incorporated in the active layers or between the layers or both should be further studied.

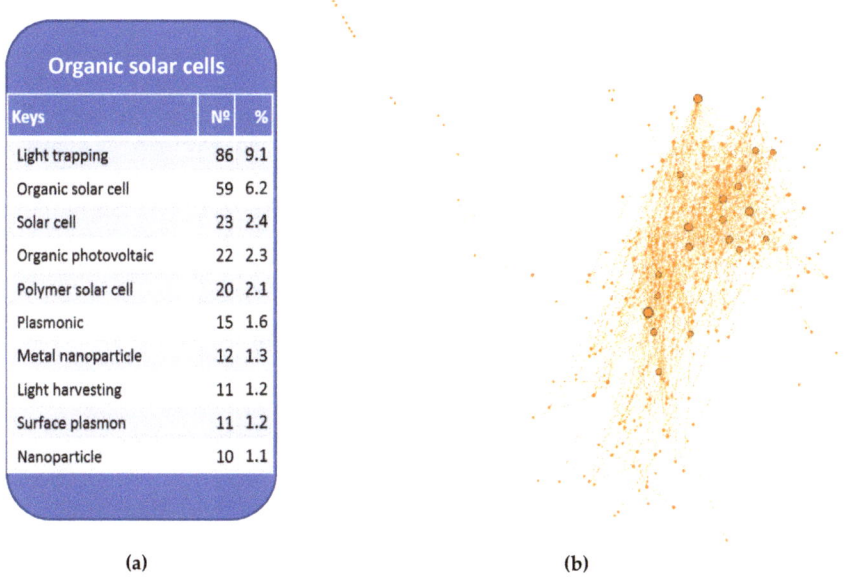

Figure 6. Representation of Organic solar cells community: (a) keywords; and, (b) isolated distribution of the publications.

The review that was carried out by Gan et al. [43] proposes incorporating plasmonic nanostructures in the front and rear metal electrodes of an OSC, which is expected to reach broadband, polarization, and absorption independent of the angle, and this implies the possibility of exceeding 10% power conversion efficiency. Tvingstedt et al. [44] analyzes the use of micro lens to increase the capture of light and, thereby, improves the absorption rate of the solar cell. Xiao et al. [45] propose a hybrid system of micro lens for OSC (a matrix of hybrid micro lens, a mirror with a matrix of holes, and an OSC with a reflective cathode) to improve broadband absorption. Each isana chromatic hybrid refractive-diffractive singlet micro lens made of a single optical material, and these hybrids micro lens are separated from the cells to avoid direct contact with an organic layer that can cause electrical defects. Another proposal is the use of a V-shaped light capture configuration; the purpose is to increase the photocurrent for all angles of incidence. Rim et al. [46] tested in a 170 nm polymer thin film OSC, obtaining a 52% improvement in efficiency and conclude that this V structure in thin film OSC is effective for active layer thicknesses of the order of wavelength of light or less.

Müller-Meskamp et al. [47] study direct patterning interfering laser (DLIP) has been used to manufacture periodic surface patterns (substrates with a 4.7 μm and hexagonal line of 0.7 μm) large surface area on flexible polyethylene terephthalate (PET) substrates. The results are encouraging, achieving the best results for the hexagonal corrugated structures with greater short-circuit current (Jsc) and greater energy conversion efficiency (PCE). Other important studies should be cited for Perovskite Solar Cells related to High-Performance Solution-Processed Double-Walled Carbon Nanotube Transparent Electrode [48], and the Highly reproducible perovskite solar cells with an average efficiency of 18.3% and best efficiency of 19.7% being fabricated via Lewis base adduct of lead (II) iodide [49]. Publications of the most cited in Scopus in this cluster are:

- "Plasmonic-enhanced organic photovoltaics: Breaking the 10% efficiency barrier" [50]. Cited 391 times.
- "An effective light trapping configuration for thin-film solar cells" [46]. Cited 170 times.
- "Light manipulation in organic photovoltaics" [51]. Cited 23 times.

Community 3 (Thin film a-Si: H) presents advances on thin-layer solar cells of hydrogenated Amorphous (a-Si:H) or hydrogenated microcrystalline (μc-Si:H). Figure 7a shows the most representative keywords of the Thin film a-Si: H community, which shows the number of times and their percentage of repetition within the community. The most representative word is "Thin film solar cells", which makes part of the community name, the next word is "Silicon", which is generic, and the rest are already more specific to the topic treated in this community, such as " Amorphous silicon "and" Microcrystalline silicon ". This community, although its studies are focused on a-Si cells: H is the second community in relation to total publications (Figure 2). It has a central core with a large number of references and three somewhat lower, but considerable references (Figure 7b). In addition, it maintains a great interaction with the rest of the communities that supply it with references. The publications of the most referenced nodes in Scopus in the order of size are the following three:

- "TCO and light trapping in silicon thin film solar cells" [52] with 869 cites.
- "Light trapping in solar cells: Can periodic beat random?" [53] was cited 369 but its node is large because it relates to major nodes.
- "Light trapping in ultrathin plasmonic solar cells" [54] had 512 cites, but with lower relations to major nodes.

The thin-layered Si (H-Si:H) or hydrogenated microcrystalline (μc-Si:H) thin-layer solar cells use an intrinsic layer (layer i) without doping between two highly doped layers (p and n). The optical and electrical properties of the i-layers are linked to the microstructure and, therefore, to the deposition rate of the layer i, which in turn affects the production yield [55]. The importance of contact and reflection in these solar cells require techniques to improve light uptake [52]. An integral part of these devices is the transparent conductor oxide (TCO) layers used as a front electrode and as a part of the rear side reflector [56]. When applied on the front side, the TCO must have high transparency in the spectral region, where the solar cell operates with high electrical conductivity. In p-i-n configuration, where the Si layers are deposited on a transparent substrate covered by TCO, with rough surfaces are applied in combination with the highly reflective rear contacts. TCO must have a strong dispersion of the incoming light in the silicon absorbent layer and favorable physicochemical properties for silicon growth. The application of zinc oxide films that were doped with aluminum (ZnO:Al) as a rear reflector result in a highly promising TCO material [57]. These films provide efficient coupling of the incident sunlight by refraction and light scattering at the interface TCO/Si to increase the length of the light path [52]. Another proposal for improving light entrapment is to use a return reflector; the use of Ag plasmonic nanoparticles can provide performance that is comparable to random textures in amorphous silicon solar cells n-i-p [58].

Thin film a-Si:H		
Keys	№	%
Thin film solar cells	97	3.0
Silicon	40	1.2
Amorphous silicon	39	1.2
Microcrystalline silicon	32	1.0
Back reflector	31	1.0
ZnO	28	0.9
Light scattering	26	0.8
Photovoltaics	26	0.8
Zinc oxide	25	0.8
Plasmonics	20	0.6

(a)   (b)

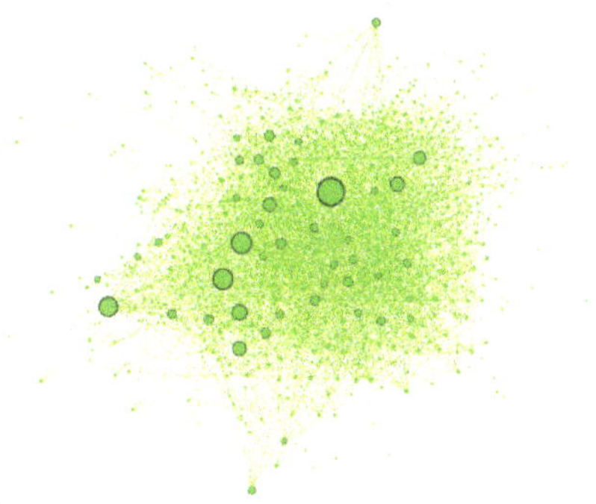

Figure 7. Representation of Thin film Si:H community: (**a**) keywords; and, (**b**) isolated distribution of the publications.

Other researchers propose different nanostructures for improving light entrapment, which are very important in thin-film amorphous silicon solar cells. Battaglia et al. [53] compare a random pyramidal nanostructure of transparent zinc oxide electrodes and a periodic one of periodic glass nanocavity matrixes manufactured by nanosphere lithography. The results show that both options have approximately a short-circuit current density of 17.1 mA/cm^2 and a high initial efficiency of 10.9%. Waveguide Theory provides a mechanism to select the period and symmetry of the grid to obtain efficiency improvements. The relationship between the photocurrent and the spatial correlations of random surfaces have been proposed by [59], developing pseudo-random matrices of nanostructures that are based on their power spectral density, and their correlation between the frequencies and the photocurrent.

Another option is nanodome solar cells, which have periodic nanoscale modulation for all types of solar cells from the lower substrate, through the active absorber to the upper transparent contact. These devices combine many nanophotonic effects to efficiently reduce the reflection and improve absorption over a wide spectral range. Nanodome solar cells with only one layer of 280 nm thick hydrogenated amorphous silicon (a-Si: H) can absorb 94% of the light with wavelengths of 400–800 nm, which is significantly greater than 65% absorption of flat film devices. In [60], they propose a nanodome solar cell of union p-i-n a-Si: H. The cells are composed of 100 nm thick Ag as a rear reflector, 80 nm Thick transparent conduction oxide (TCO) as a bottom and upper electrode, and a thin active layer of a-Si: 280 nm H (top to bottom): pin, 10-250-20 nm). Ferry et al. [54] proposed a strategy that consists on the use of non-randomized nanostructured reflectors optimized for ultra-thin solar cells of hydrogenated amorphous Si (a-Si:H). This alternative increases the short-circuit current densities, which improves the results as compared to cells that have posterior contacts with a flat or random texture.

Community 4 (Optical design) works on improvements in the efficiency of solar cells from the point of view of optical design by creating nanostructures to improve the capture of direct and diffused light. Figure 8a shows the most representative keywords of the optical design community, which shows the number of times and their percentage of repetition within the community. All of the words are very representative of the community, such as "Diffractive Optics" or "Optical Design", after which the community is named and it represents 11.5% each with respect to the community; the next words are "Diffractive Optics elements" and "Diffractive elements." This is the second smallest community, although it maintains links with the rest of the communities (Figure 2). This community incorporates publications from other disciplines, such as optics, which were not initially developed for solar energy, but whose impact on the optical behavior of a surface has been referenced by the other communities. In Figure 2, it appears as an emerging community. Unlike the other communities, it does not have a main nucleus, since this community does not have a great concentration of relationships in its articles (Figure 8b).

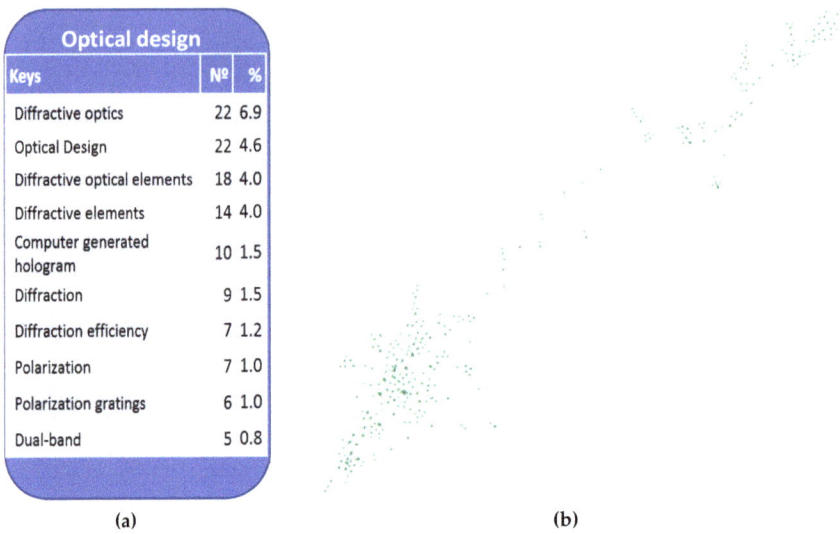

Figure 8. Representation of Optical design community: (a) keywords; and, (b) isolated distribution of the publications.

The manufacture of solar cells requires a prior study of their optical behavior with the consideration of better light capture. Therefore, their behavior is studied as a diffractive element, and lithographic structures respond differently, depending on their structure, composition, and size. Herkommer et al. [61] show simulation techniques to evaluate the distraction efficiency prior to the manufacture of the solar cell. It is very important prior to manufacturing to simulate the behavior in real conditions, since different manufacturing systems can be used, depending on the wavelength range of the incident and its size characteristics. Depending on the wavelength, more than one material can be used to implement the grid structures. The dependence on the size of the diffraction efficiency characteristic can be considered as the manufacturing limitations. There is commercial software that calculates the diffraction efficiency for a given grid while considering the period, grid shape, wavelength, and angle of incidence.

The diffractive optical element (DOE) can be implemented on flat and curved surfaces [12]. Grilles can be designed and implemented more complexly according to the need for the range of light collection and its efficiency with the use of direct laser lithography or through computer-generated holograms. On curved surfaces, they can be manufactured while using single-point diamond turning [62] to reduce the material used.

Digital holography (computer-generated hologram) has improved its quality in recent years, leading to unthinkable implementations for its accuracy not many years ago. Patterns can be engraved on photosensitive materials at the appropriate scale, embossed on high precision materials in various directions with or without periodicity, with the arrival of the laser and its computer control. Some lithographic techniques combine the coating with a light sensitive film (photo-resistance). Digital holography and Computer-Generated Holograms reduce the choice of material and pattern generation scheme. The degree of freedom in the choice of parameters limits the choice of coding technique and its optimization. The coding allows for adapting the data to the existing hardware requirements [12].

Solar cells have benefited from the study of the behavior of diffractive elements applied to other fields of research. The most common diffractive elements are a diffractive lens, a matrix generator, and a correlation filter [63]. The design is based on the optimum performance of the optical system and its manufacturing restrictions. These diffractive elements have a disadvantage in that they produce chromatic aberrations. Some authors [64] propose the attachment to the lens of a corrective substrate of the diffraction for application in the headlights of a car and study these elements for the diffractive telescope system.

Hybrids are one of the most complex diffractive elements, where they are both reflective and dissipative throughout the visible band (400–700 nm). Designing achromatic refractive-diffraction hybrid lenses is complex and it requires prior study for its manufacture to optimize its efficiency in the capture range [65]. Publications regarding the most cited in Scopus in this cluster are:

- "Digital holography as part of diffractive optics" [66] cited 111 times.
- "Understanding diffractive optic design in the scalar domain" [67] cited 110 times.
- "I digital holography computer-generated holograms" [68] cited 23 times.

Community 5 (Light trapping) investigates the improvement of the capture of light in solar cells, while using nanoparticles for the purpose of improving efficiency. Although many authors investigate and test systems for improving the light entrapment by nanostructures, in their turn they produce an increase in surface area and, therefore, minority recombination that reduces the efficiency. Other researchers use the dispersion of metal nanoparticles by varying their shape, size, particle material, and ambient dielectric energy to determine the improvement of light capture with particle plasmons to avoid this adverse situation. Figure 9a shows the most representative keywords of the Light trapping community, which shows the number of times and their percentage of repetition within the community. All of the words are very representative of the community as "Plasmonic" or "Surface plasmons", and they represent 4% and 3.1%, each with respect to the community, other words, like "Nanoparticles", "Silver nanoparticles", and "Metal nanoparticles", are very specific to the community, although within the 10 most repeated words, there are other generics, such as "Photovoltaics" and "Silicon" with 2.5% and 1.6% each. This community is ranked third in size (Figure 2) and it has a main nucleus and a somewhat smaller one as seen in Figure 9b. It is a community that is closely related to the rest, since the main topic discussed affects the studies of other communities, such as the entrapment of light, which, regardless of the type of cell, is important to improve and increase the efficiency. The two publications of the most referenced nodes in Scopus in the order of size are the following:

- "Surface plasmon enhanced silicon solar cells" with 1467 references and many relations with other communities [69].
- "Design principles for particle plasmon enhanced solar cells" [70] with 673 cites.
- "Tunable light trapping for solar cells using localized surface plasmons" [71] with 460 cites.

Catchpole et al. [70] show that the shape of the particles influences the path length, with the spherical shapes being worse than the cylindrical and hemispherical shapes. In addition, they conduct experiments with silver and gold particles, where the results show that those of silver provide a longer path length than those of gold. In Ouyang et al. [9], the effects of silver nanoparticles on polycrystalline silicon thin film solar cells on glass are studied, obtaining an improvement in the short-circuit current of a 1/3 increase when compared to conventional ones. The geometry of the matrix used also defines the entrapment of light (there are studies on random, quasi-periodic, and periodic matrices). Mokkapati et al. [72] study the behavior of silver nanoparticles in a periodic matrix and conclude that there is a very restrictive relationship between the optimal particle size and grid parameters of the periodic matrix; the case of silver particles of 200 nm, a 400 nm step is ideal for Si solar cells. Other authors analyzed the behavior of silver nanoparticles on the rear structure to reduce the entrapment losses that were produced with the long wavelength that escape (rear reflector dispersion), but with minimal electronic losses due to recombination effects. The photocurrent with the silver nanoparticles of a PERT (Passivated Emitter and Rear Totally Diffused) cell increases by 16% when compared to an aluminum rear structure [73].

Light trapping		
Keys	Nº	%
Light trapping	104	7.9
Solar cells	65	5.0
Plasmonics	53	4.0
Surface plasmons	41	3.1
Photovoltaics	33	2.5
Nanoparticles	28	2.1
Silver nanoparticles	23	1.8
Silicon	21	1.6
Thin film solar cells	21	1.6
Metal nanoparticles	11	0.8

(a)

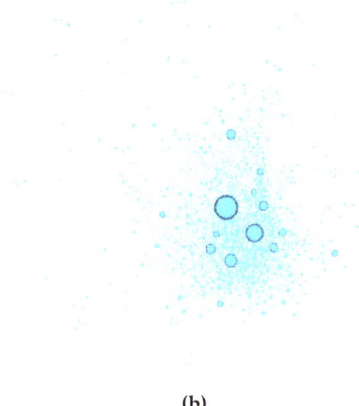

(b)

Figure 9. Representation of Light trapping community: (**a**) keywords, and (**b**) isolated distribution of the publications.

3.3. Analysis as Per Authors, Affiliations, and Countries of Investigation on Solar Coating

In total, there are 14,849 authors who research in 26 by subject area. The ten countries with the highest concentration of researchers contributing their scientific publications to the progress in Solar Coating have been analyzed. China makes the main contribution, with 3345 researchers (22.5%), being closely followed by the United States with 2634 (17.7%) and Germany with 1156 (7.8%). The rest is around 349 from Italy to 687 in South Korea, with percentages ranging from 2.4 to 4.6%, respectively. China and the United States are among the countries with the greatest contributions in scientific developments in all matters related to energy, and also in energy saving [16]. Figure 10 shows the distribution and percentages of participation of the authors according to the country of origin shown by colors. There are 28.8% of other countries of low percentage contribution, as can be seen in Figure 10, where Spain, Australia, and others appears.

The results show that China and the United States both actively collaborate in the progress by establishing collaborative ties with other countries in order to move forward and this can be seen from the intertwining of the communities in Figure 2. The main collaborations are established in Asia between Chinese, Koreans, Taiwanese, Japanese, and Indians, making a total of 38%. The European contributions between Germany, France, United Kingdom, and Italy make up for 16%. America is only represented by the United States.

Going deeper into the origin of the authors, the 20 most localized affiliations in the search have been analyzed, since it is important to know in which research and development centers the main contributions to Solar Coating are made. Table 1 shows the data; 12 of the 20 belong to the Asian continent with a contribution of 65% of the Top 20 publications. It is interesting how Switzerland, which was not among the 10 countries with the highest productivity in Solar Coating, is the second research institution in terms of the number of contributions with 6.6% of the top 20. In Europe, Germany stands out with the centers of the Fraunhofer Institute for Solar Energy Systems ISE and Helmholtz-Zentrum Berlin für Materialien und Energie (HZB) with a total of 11.1%, which together with Switzerland contribute in 17.7%. America contributes in 12.6% with the USA and only three research institutions, being the one that Stanford University publishes the most.

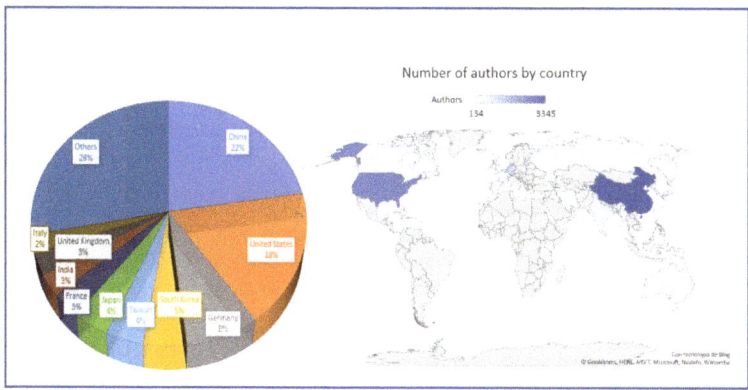

Figure 10. Author distribution per countries.

Table 1. Top 20 affiliations in Solar Coating.

Affiliation	Country	Publications
National Taiwan University	Taiwan	123
Swiss Federal Institute of Technology, Lausanne	Switzerland	115
Fraunhofer Institute for Solar Energy Systems ISE	Germany	108
Shanghai Jiao Tong University	China	105
Nankai University	China	103
Soochow University	China	103
Chinese Academy of Sciences	China	99
University of New South Wales (UNSW)	Australia	91
Forschungs zentrum jülich (FZJ)	Germany	84
Helmholtz-Zentrum Berlin für Materialien und Energie (HZB)	Germany	83
Stanford University	USA	82
Nanjing University	China	79
National Chiao Tung University	China	77
Jilin University	China	76
Sungkyunkwan University	South Korea	74
Nanyang Technological University	Singapore	69
National Renewable Energy Laboratory	USA	69
Massachusetts Institute of Technology	USA	67
Australian National University	Australia	64
Sun Yat-Sen University	China	62

Table 2 shows a list of the 10 authors with the highest H-index in Scopus in relation to the search performed. The order of the authors does not correspond to the order by institution of Table 1. The author with the highest H-index of 230, 1471 published documents, and 258,107 citations belong to Switzerland and there is a second in position 9 both same affiliations appearing in Table 1 (Swiss Federal Institute of Technology, Lausanne). In the top 10 by H-index, there are mainly US authors from Harvard University, Georgia Institute of Technology, Stanford Linear Accelerator Center, and University of California (Berkeley); none correspond to the three affiliations in Table 1. There are two Chinese authors in positions 8 and 10, with H-indexes of 138 and 130, respectively, which have a number of significant citations of 77,076 and 72789, in both cases their center does not correspond to those of Table 1. There is a German author with H-index of 186 and 136,496 citations, but their affiliation does not correspond to the affiliations of Germany in Table 1. It is significant that only the affiliation of the authors of Switzerland corresponds within those seen in Table 1. This might be due to the fact that, although the group of contributions has great references, it does not establish collaborations between the communities, and that means that the software debugging process has been eliminated. Figure 11 shows the location of the author of the highest H-index in Table 2. As it can be seen in the figure, it corresponds to the border that has not been represented when analyzing the communities, since it remains as a node with no connection to the rest. This can be interpreted that the system correctly measures relationships and not just references, giving more value to relationships.

Table 2. Top 10 authors by H-index (Nco-author = Number of coauthors, Ncite = Number of cites, Ndoc = Number of documents).

Author Scopus ID	Name	H-Index	Nco-Author	Ncite	Ndoc	City	Country	Affiliation
35463345800	Gratzel M.	230	2222	258107	1471	Lausanne	Switzerland	Swiss Federal Institute of Technology, Lausanne
55711979600	Whitesides G.	187	1268	159450	1008	Cambridge	United States	Harvard University
7403027697	Xia Y.	186	952	136496	779	Atlanta	United States	Georgia Institute of Technology
7103185149	Antonietti M.	156	1047	84786	789	Golm	Germany	Max Planck Institut fur Kolloid Und Grenzflächenforschung Potsdam
35207974600	Cui Y.	155	994	99093	532	Menlo Park	United States	Stanford Linear Accelerator Center
7403931988	Yang P.	150	706	102859	408	Berkeley	United States	University of California, Berkeley
56605567400	Alivisatos A.	143	934	108954	445	Berkeley	United States	University of California, Berkeley
7403489871	Zhao D.	138	1189	77076	685	Shanghai	China	Fudan University
35463772200	Nazeeruddin M.	133	1090	82713	576	Lausanne	Switzerland	Swiss Federal Institute of Technology, Lausanne
56422845100	Jiang L.	130	2113	72789	1256	Beijing	China	Technical Institute of Physics and Chemistry Chinese Academy of Sciences

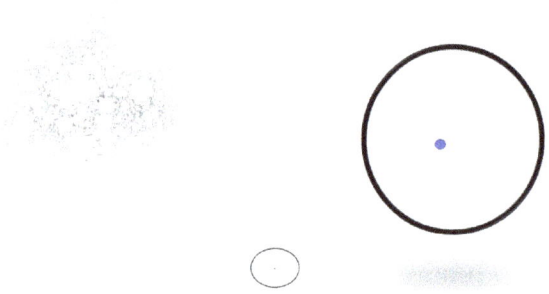

Figure 11. Location of the Author with the highest H-index with research in Solar Coating.

If we analyze the top 10 of the authors considered in the analyzed communities (Table 3), it is observed that they have lower H-index than those shown in Table 2 and, in this case, if they correspond to the affiliations in Table 1. An author from Belgium appears and another from France, which does not appear from its affiliations in Table 2, despite being in positions 5 and 8 of Table 3.

Table 3. Most important 10 for H-index analyzed in the communities (Nco-author = Number of coauthors, Ncite = Number of cites, Ndoc = Number of documents).

Author Scopus ID	Name	H-Index	Nco-Author	Ncite	Ndoc	City	Country	Affiliation
6701805412	Ballif C.	65	750	13344	461	Lausanne	Switzerland	Swiss Federal Institute of Technology, Lausanne
56216991600	Rech B.	48	514	10525	321	Berlin	Germany	Helmholtz-Zentrum Berlin für Materialien und Energie (HZB)
6603242760	Blasi B.	21	201	1542	105	Freiburg im Breisgau	Germany	Fraunhofer Institute for Solar Energy Systems ISE
7004315658	Haug F.	37	282	5174	171	Lausanne	Switzerland	Swiss Federal Institute of Technology, Lausanne
56597035200	Poortmans J.	44	781	8334	523	Leuven	Belgium	Interuniversity Micro-Electronics Center at Leuven
55505896100	Zhao Y.	21	338	2607	462	Tianjin	China	Nankai University
55931076700	Zhang X.	17	420	2020	383	Tianjin	China	Nankai University
6602741595	Roca i Cabarrocas P.	43	683	7634	497	Palaiseau	France	Laboratoire de Physique des Interfaces et des Couches Minces
7402736621	Yi J.	33	617	4594	471	Jongno-gu	South Korea	Sungkyunkwan University
7006823424	Atwater H.	89	726	41733	654	Pasadena	United States	California Institute of Technology

4. Discussion

The scientific contributions that were published in the Solar Coating field from 1964 to June 2019 make a total of 6440 and 127 new ones published as of September. Until 1982, the publications were sporadic with a couple of publications per year, and then increased in 1990, but did not exceed 20 annual publications. From then on, the contributions are more numerous until 2004, being marked with 97 publications. From 2005 to 2019, this period is marked with a high concentration of publications (83%), due to the "renewable boom". The fact that the highest concentration has occurred, since 2005 might be due to global awareness of environmental concerns, such as global warming and the greenhouse effect, as shown by the agreement in Kyoto, Japan, on 11 December 1997 and entered into force on February 16, 2005. This has meant that the scientific community has turned to providing solutions; one of the clean energies that is most committed to the future is photovoltaic solar energy [73]. The main forms of publications are in original articles with 62.6% and in conferences with 32.6%, followed by far by review articles and book chapters with 1.9% and 1.1%, respectively. Book contributions are scarce, with only six published < 0.01% [74–80] and all with little impact on the number of references below 22. In [16], the authors analyze the bibliometric in energy saving and obtain a percentage of publication of articles of 50.7%, conferences 43.1%, and in books a percentage lower than 1%, which is very much in line with those that were obtained in this work.

English is the most used language for publications with 96.8% of the total publications, followed by Chinese, with a percentage of 2.6%, and the rest of the languages have a representation lower than < 1% (Russian, Japanese, Korean, German, Spanish, French, Lithuanian, Finnish, and Malay). It is usual for the English language to be the most widespread in scientific publications due to the edition standards of the journals to have maximum dissemination, regardless of the country of origin of the headquarters of the publisher. Only China produces a number of scientific publications in its own language, such as Taiyangneng Xuebao/Acta Energiae Solaris Sinica, Cailiao Gongcheng/Journal of Materials Engineering among others. The demand for energy increases with population growth and China is one of the largest consumers, due to its industrial expansion and population growth, which motivates China to investigate the production of clean energy, as seen in the results shown in Tables 1–3.

The thematic areas that are most used for scientific dissemination are mainly Physics and Astronomy (25.9%), Materials Science (23.6%), Engineering (21.6%), Energy 6.4%), Computer Science (5.6%), Chemistry, Mathematics (4.6%), Chemical Engineering (4.6%), and the rest with a contribution of less than 3%. The results show the great involvement of researchers from different disciplines, such as engineering, materials sciences, chemistry, and mathematics, among others, having all the purpose of contributing to improvements in solar energy collection. The contributions of the application of nanomaterials and nanostructures to the collection surfaces have allowed for improving efficiency by trapping light and generating new flexible, bifacial panels that have expanded the use of these panels as an architectural tool in industrial buildings and facilities, electric locomotion, and solar powered devices, thereby reducing the use of fossil fuels and CO_2 emissions.

5. Conclusions

This work has revealed data related to Solar Coating from 1957 to 2019. The total contributions found reached 6440 documents with a total of 21,301 relations between authors, where only 3924 with 20,849 relations after the use of bibliometric techniques. The study of efficiency improvements has driven a large number of contributions in different sub-themes, where each community develops its research on different perspectives, although they all have the improvement of light trapping as a system for improving efficiency in common. Publications have focused on six categories or communities Material Solar Cell, Thin Film and Polycrystalline, Organic solar cells Thin film a-Si: H, Optical design, and Light trapping. The communities with the highest representation are the most generic: The Material Solar Cell (42%), Thin film a-Si: H (25%), and Light trapping (11%). However, from a different perspective study aspect, they are provided with a main nucleus and multiple relations with the rest of the communities that nourish them of references. Organic solar cells (8.3%), Optical design (8%), and Thin Film and Polycrystalline (6%) are the other three communities. In addition to having less representativeness, these communities do not have a main nucleus, but small nuclei with a multitude of relationships in and between communities, with organic solar cells having 19 major nuclei, although much lower than all of the main ones of the more generic communities. The most repeated keywords correspond to the generic communities, which is consistent because they are the ones with the greatest ties and the most referenced publications. There are communities, like Optical design with keywords, which are very specific to their theme. This is because this community is tangential to Solar Coating incorporating research on the theme of Optics that are not specifically developed for solar cells, but their advances substantially influence the improvements of light capture, which is an essential theme for Solar Coating.

In total, there are 14,849 authors who research in 26 by subject area, China (3345 researchers) and the United States (2634 researchers) are from the countries with the greatest contributions in scientific developments in all matters related to energy, also in energy saving. The worldwide distribution is established between Asia (38%), Europe (16%), and America, with only the United States (17.7%), and there is no significant representation of the African continent. If we consider the authors' affiliation with a top 20, 12 of the 20 belong to the Asian continent (65%). It is interesting how Switzerland, which did not appear among the 10 countries with the highest productivity in Solar Coating, is the second research institution in terms of the number of contributions with 6.6% of the top 20. In Europe, Germany stands out with the Fraunhofer Institute for Solar Energy Systems centers ISE and Helmholtz-Zentrum Berlin für Materialien und Energie (HZB) (11.1%). In the USA, which has 18% of the worldwide publications, with respect to the ranking of the top 20 affiliations only three research institutions are present, from them the best positioned is Stanford University.

The analysis of the 10 authors with the highest H-index in Scopus in relation to the search performed does not correspond to the order by institution, for example, the author with the highest H index of 230, 1471 published documents, and 258,107 citations belongs to the Swiss Federal affiliation Institute of Technology of Lausanne, (Switzerland). On the other hand, the first author in the Top 20 by affiliation is from National Taiwan University (Taiwan).

The main forms of publications are in original articles (62.6%), conferences (32.6%), followed by review articles (1.9%) and book chapters (1.1%), with scarce book contributions (<0.01%). As for the language, English (96.8%) is the most used for disseminating publications due to the standards of edition of the journals to have maximum dissemination, regardless of the country of origin of the headquarters of the publisher. The second language used is Chinese (2.6%), as it produces a significant number of scientific publications with international repercussions in their own language. All other languages have a representation of less than <1%.

The most used thematic areas are very diverse; this shows the great involvement of researchers from different disciplines, such as engineering, materials sciences, chemistry and mathematics, among others, all with the purpose of contributing to improvements in solar energy collection.

The final conclusion of the present work shows that the research on this subject is not completed and it requires even more research with different considerations that improve solar cell efficiencies. The joint collaboration of all researchers is required, with their valuable contributions opening new perspectives regarding the necessary improvement due to the growing need for global demand for energy consumption and the awareness of consumers and political entities of environmental care and implementation of renewable energy.

Author Contributions: N.N. formed the manuscript; A.A. and F.M.-A. developed the figures and tables; A.A. and D.E.K. contributed to the search of date and the realized of the maps; F.M.-A. and N.N. redacted the paper. D.E.K. checked the whole manuscript.

Funding: Under I+D+I Project—FEDER Andalucía 2014-2020 Operational Programme, UAL18-TIC-A025-A.

Acknowledgments: The Ministry of Economic and Competitiveness of Spain financed this work, under Project TEC2014-60132-P, in part by Innovation, Science and Enterprise, Andalusian Regional Government through the Electronics, Communications and Telemedicine TIC019 Research Group of the University of Almeria, Spain and in part by the European Union FEDER Program and CIAMBITAL Group.

Conflicts of Interest: The authors declare no conflict of interest.

References

1. McCollum, D.; Gomez Echeverri, L.; Riahi, K.; Parkinson, S. Affordable and clean energy: Ensure access to affordable, reliable, sustainable, and modern energy for all. In *Atlas of Sustainable Development Goals 2018: From World Development Indicators*; International Bank for Reconstruction and Development/The World Bank: Washington, WA, USA, 2018; pp. 26–29.
2. Hansen, K.; Vad Mathiesen, B. Comprehensive assessment of the role and potential for solar thermal in future energy systems. *Sol. Energy* **2018**, *169*, 144–152. [CrossRef]
3. Jia, T.; Dai, Y.; Wang, R. Refining energy sources in winemaking industry by using solar energy as alternatives for fossil fuels: A review and perspective. *Renew. Sustain. Energy Rev.* **2018**, *88*, 278–296. [CrossRef]
4. Rogelj, J.; Den Elzen, M.; Höhne, N.; Fransen, T.; Fekete, H.; Winkler, H.; Meinshausen, M. Paris Agreement climate proposals need a boost to keep warming well below 2 °C. *Nature* **2016**, *534*, 631–639. [CrossRef] [PubMed]
5. Kilinc-Ata, N. The evaluation of renewable energy policies across EU countries and US states: An econometric approach. *Energy Sustain. Dev.* **2016**, *31*, 83–90. [CrossRef]
6. Jiaru, H.; Xiangzhao, F. An evaluation of China's carbon emission reduction policies on urban transport system. *J. Sustain. Dev. Law Policy* **2016**, *6*, 31. [CrossRef]
7. Vasiliev, M.; Nur-e-alam, M.; Alameh, K. Initial field testing results from building-integrated solar energy harvesting windows installation in Perth, Australia. *Appl. Sci.* **2019**, *9*, 4002. [CrossRef]
8. Catchpole, K.R.; Mokkapati, S.; Beck, F.; Wang, E.C.; McKinley, A.; Basch, A.; Lee, J. Plasmonics and nanophotonics for photovoltaics. *MRS Bull.* **2011**, *36*, 461–467. [CrossRef]
9. Ouyang, Z.; Pillai, S.; Beck, F.; Kunz, O.; Varlamov, S.; Catchpole, K.R.; Green, M.A. Effective light trapping in polycrystalline silicon thin-film solar cells by means of rear localized surface plasmons. *Appl. Phys. Lett.* **2010**, *96*, 261109. [CrossRef]
10. Rand, J.A.; Basore, P.A. Light-trapping silicon solar cells-experimental results and analysis. In Proceedings of the Conference Record of the IEEE Photovoltaic Specialists Conference, Las Vegas, NV, USA, 7–11 October 1991; pp. 192–197.
11. Bokor, N.; Davidson, N. Curved diffractive optical elements: Design and applications. *Prog. Opt.* **2005**, *48*, 107–148.
12. Cirino, G.A.; Verdonck, P.; Mansano, R.D.; Pizolato, J.C., Jr.; Mazulquim, D.B.; Neto, L.G. Digital holography: Computer-generated holograms and diffractive optics in scalar diffraction domain. In *Holography-Different Fields of Application*; Monroy, F., Ed.; IntechOpen: London, UK, 2011. [CrossRef]

13. Luo, M.; Zhou, L.; Yuan, J.; Zhu, C.; Cai, F.; Hai, J.; Zou, Y. A new non-fullerene acceptor based on the heptacyclic benzotriazole unit for efficient organic solar cells. *J. Energy Chem.* **2020**, *42*, 169–173. [CrossRef]
14. Montoya, F.G.; Alcayde, A.; Baños, R.; Manzano-agugliaro, F. Telematics and Informatics A fast method for identifying worldwide scientific collaborations using the Scopus database. *Telemat. Inf.* **2018**, *35*, 168–185. [CrossRef]
15. Mongeon, P.; Paul-Hus, A. The journal coverage of Web of Science and Scopus: A comparative analysis. *Scientometrics* **2016**, *106*, 213–228. [CrossRef]
16. Cruz-Lovera, C.; Perea-Moreno, A.-J.; de la Cruz-Fernández, J.L.; Montoya, F.G.; Alcayde, A.; Manzano-Agugliaro, F. Analysis of Research Topics and Scientific Collaborations in Energy Saving Using Bibliometric Techniques and Community Detection. *Energies* **2019**, *12*, 2030. [CrossRef]
17. Jacomy, M.; Venturini, T.; Heymann, S.; Bastian, M. ForceAtlas2, a continuous graph layout algorithm for handy network visualization designed for the Gephi software. *PLoS ONE* **2014**, *9*, e98679. [CrossRef] [PubMed]
18. Garnett, E.; Yang, P. Light trapping in silicon nanowire solar cells. *Nano Lett.* **2010**, *10*, 1082–1087. [CrossRef] [PubMed]
19. Yu, Z.; Raman, A.; Fan, S. Fundamental limit of nanophotonic light trapping in solar cells. *Next Gener. Photonic Cell Technol. Sol. Energy Convers.* **2010**, *7772*, 77720Z.
20. Bermel, P.; Luo, C.; Zeng, L.; Kimerling, L.C.; Joannopoulos, J.D. Improving thin-film crystalline silicon solar cell efficiencies with photonic crystals. *Opt. Express* **2007**, *15*, 16986–17000. [CrossRef]
21. Kelzenberg, M.D.; Boettcher, S.W.; Petykiewicz, J.A.; Turner-Evans, D.B.; Putnam, M.C.; Warren, E.L.; Atwater, H.A. Enhanced absorption and carrier collection in Si wire arrays for photovoltaic applications. *Nat. Mater.* **2010**, *9*, 239–244. [CrossRef]
22. Mokkapati, S.; Catchpole, K.R. Nanophotonic light trapping in solar cells. *J. Appl. Phys.* **2012**, *112*, 10. [CrossRef]
23. Brongersma, M.L.; Cui, Y.; Fan, S. Light management for photovoltaics using high-index nanostructures. *Nat. Mater.* **2014**, *13*, 451–460. [CrossRef]
24. Yu, Z.; Raman, A.; Fan, S. Fundamental limit of light trapping in grating structures. *Opt. Express* **2010**, *18*, A366–A380. [CrossRef] [PubMed]
25. Yang, W.; Yu, H.; Wang, Y. Light transmission through periodic sub-wavelength textured surface. *J. Opt.* **2013**, *15*, 5. [CrossRef]
26. Wang, K.X.; Yu, Z.; Liu, V.; Cui, Y.; Fan, S. Absorption enhancement in ultrathin crystalline silicon solar cells with antireflection and light-trapping nanocone gratings. *Nano Lett.* **2012**, *12*, 1616–1619. [CrossRef]
27. Aberle, A.G.; Widenborg, P.I. Crystalline Silicon Thin-Film Solar Cells Via High-Temperature and Intermediate-Temperature Approaches. In *Handbook of Photovoltaic Science and Engineering*, 2nd ed.; Wiley: New York, NY, USA, 2011.
28. Eisenhauer, D.; Trinh, C.T.; Amkreutz, D.; Becker, C. Light management in crystalline silicon thin-film solar cells with imprint-textured glass superstrate. *Sol. Energy Mater. Sol. Cells* **2019**, *200*, 109928. [CrossRef]
29. Xue, C.; Rao, J.; Varlamov, S. A novel silicon nanostructure with effective light trapping for polycrystalline silicon thin film solar cells by means of metal-assisted wet chemical etching. *Phys. Stat. Solidi Appl. Mater. Sci.* **2013**, *210*, 2588–2591. [CrossRef]
30. Gall, S.; Becker, C.; Conrad, E.; Dogan, P.; Fenske, F.; Gorka, B.; Rech, B. Polycrystalline silicon thin-film solar cells on glass. *Sol. Energy Mater. Sol. Cells* **2009**, *93*, 1004–1008. [CrossRef]
31. Mehmood, H.; Tauqeer, T.; Hussain, S. Recent progress in silicon-based solid-state solar cells. *Int. J. Electron.* **2018**, *105*, 1568–1582. [CrossRef]
32. Jia, G.; Andrä, G.; Gawlik, A.; Schönherr, S.; Plentz, J.; Eisenhawer, B.; Falk, F. Nanotechnology enhanced solar cells prepared on laser-crystallized polycrystalline thin films (<10 µm). *Sol. Energy Mater. Sol. Cells* **2014**, *126*, 62–67.
33. Xiao, S.; Wang, T.; Liu, Y.; Xu, C.; Han, X.; Yan, X. Tunable Light Trapping and Absorption Enhancement with Graphene Ring Arrays. *Phys. Chem. Chem. Phys.* **2016**, *18*, 26661–26669. [CrossRef]
34. Abenante, L. Optical path length factor at near-bandgap wavelengths in Si solar cells. *IEEE Trans. Electron Devices* **2006**, *53*, 3047–3053. [CrossRef]
35. Yablonovitch, E.; Cody, G.D. Intensity Enhancement in Textured Optical Sheets for Solar Cells. *IEEE Trans. Electron Devices* **1982**, *29*, 300–305. [CrossRef]

36. Becker, C.; Amkreutz, D.; Sontheimer, T.; Preidel, V.; Lockau, D.; Haschke, J.; Jogschies, L.; Klimm, C.; Merkel, J.J.; Plocica, P.; et al. Polycrystalline silicon thin-film solar cells: Status and perspectives. *Sol. Energy Mater. Sol. Cells* **2013**, *119*, 112–123. [CrossRef]
37. Dross, F.; Baert, K.; Bearda, T.; Deckers, J.; Depauw, V.; El Daif, O.; Gordon, I.; Gougam, A.; Govaerts, J.; Granata, S.; et al. Crystalline thin-foil silicon solar cells: Where crystalline quality meets thin-film processing. *Prog. Photovolt. Res. Appl.* **2012**, *20*, 770–784. [CrossRef]
38. Becker, C.; Preidel, V.; Amkreutz, D.; Haschke, J.; Rech, B. Double-side textured liquid phase crystallized silicon thin-film solar cells on imprinted glass. *Sol. Energy Mater. Sol. Cells* **2015**, *135*, 2–7. [CrossRef]
39. Jeong, E.; Zhao, G.; Song, M.; Yu, S.M.; Rha, J.; Shin, J.; Cho, Y.-R.; Yun, J. Simultaneous improvements in self-cleaning and light-trapping abilities of polymer substrates for flexible organic solar cells. *J. Mater. Chem. A* **2018**, *6*, 2379–2387. [CrossRef]
40. Tang, Z.; Tress, W.; Inganäs, O. Light trapping in thin film organic solar cells. *Mater. Today* **2014**, *17*, 389–396. [CrossRef]
41. Ko, D.H.; Tumbleston, J.R.; Gadisa, A.; Aryal, M.; Liu, Y.; Lopez, R.; Samulski, E.T. Light-trapping nano-structures in organic photovoltaic cells. *J. Mater. Chem.* **2011**, *21*, 16293–16303. [CrossRef]
42. Li, X.; Choy, W.C.H.; Lu, V.; Sha, W.E.I.; Ho, A.H.P. Efficiency enhancement of organic solar cells by using shape-dependent broadband plasmonic absorption in metallic nanoparticles. *Adv. Funct. Mater.* **2013**, *23*, 2728–2735. [CrossRef]
43. Feng, L.; Niu, M.; Wen, Z.; Hao, X. Recent advances of plasmonic organic solar cells: Photophysical investigations. *Polymers* **2018**, *10*, 123. [CrossRef]
44. Tvingstedt, K.; Dal Zilio, S.; Inganäs, O.; Tormen, M. Trapping light with micro lenses in thin film organic photovoltaic cells. *Opt. Express* **2008**, *16*, 21608–21615. [CrossRef]
45. Xiao, X.; Zhang, Z.; Xie, S.; Liu, Y.; Hu, D.; Du, J. Enhancing light harvesting of organic solar cells by using hybrid microlenses. *Opt. Appl.* **2015**, *45*, 89–100.
46. Rim, S.B.; Zhao, S.; Scully, S.R.; McGehee, M.D.; Peumans, P. An effective light trapping configuration for thin-film solar cells. *Appl. Phys. Lett.* **2007**, *91*, 10–13. [CrossRef]
47. Müller-Meskamp, L.; Kim, Y.H.; Roch, T.; Hofmann, S.; Scholz, R.; Eckardt, S.; Lasagni, A.F. Efficiency enhancement of organic solar cells by fabricating periodic surface textures using direct laser interference patterning. *Adv. Mater.* **2012**, *24*, 906–910. [CrossRef] [PubMed]
48. Jeon, I.; Yoon, J.; Kim, U.; Lee, C.; Xiang, R.; Shawky, A.; Xi, J.; Byeon, J.; Lee, H.M.; Choi, M.; et al. High-Performance Solution-Processed Double-Walled Carbon Nanotube Transparent Electrode for Perovskite Solar Cells. *Adv. Energy Mater.* **2019**, *9*, 1901204. [CrossRef]
49. Ahn, N.; Son, D.Y.; Jang, I.H.; Kang, S.M.; Choi, M.; Park, N.G. Highly reproducible perovskite solar cells with average efficiency of 18.3% and best efficiency of 19.7% fabricated via Lewis base adduct of lead (II) iodide. *J. Am. Chem. Soc.* **2015**, *137*, 8696–8699. [CrossRef]
50. Gan, Q.; Bartoli, F.J.; Kafafi, Z.H. Plasmonic-enhanced organic photovoltaics: Breaking the 10% efficiency barrier. *Adv. Mater.* **2013**, *25*, 2385–2396. [CrossRef]
51. Ou, Q.D.; Li, Y.Q.; Tang, J.X. Light manipulation in organic photovoltaics. *Adv. Sci.* **2016**, *3*, 1600123. [CrossRef]
52. Müller, J.; Rech, B.; Springer, J.; Vanecek, M. TCO and light trapping in silicon thin film solar cells. *Sol. Energy* **2004**, *77*, 917–930. [CrossRef]
53. Battaglia, C.; Hsu, C.M.; Söderström, K.; Escarré, J.; Haug, F.J.; Charrière, M.; Cui, Y. Light trapping in solar cells: Can periodic beat random? *ACS Nano* **2012**, *6*, 2790–2797. [CrossRef]
54. Ferry, V.E.; Verschuuren, M.A.; Li, H.B.; Verhagen, E.; Walters, R.J.; Schropp, R.E.; Polman, A. Light trapping in ultrathin plasmonic solar cells. *Opt. Express* **2010**, *18*, 237–245. [CrossRef]
55. Shah, A.V.; Schade, H.; Vanecek, M.; Meier, J.; Vallat-Sauvain, E.; Wyrsch, N.; Bailat, J. Thin-film silicon solar cell technology. *Prog. Photovolt. Res. Appl.* **2004**, *12*, 113–142. [CrossRef]
56. He, Y.; Chen, M.; Zhou, J.; Peng, T.; Ren, Z. Applying Light Trapping Structure to Solar Cells: An Overview. *Cailiao Daobao* **2018**, *32*, 696–707.
57. Berginski, M.; Hüpkes, J.; Schulte, M.; Schöpe, G.; Stiebig, H.; Rech, B.; Wuttig, M. The effect of front ZnO:Al surface texture and optical transparency on efficient light trapping in silicon thin-film solar cells. *J. Appl. Phys.* **2007**, *101*, 7. [CrossRef]

58. Tan, H.; Santbergen, R.; Smets, A.H.M.; Zeman, M. Plasmonic light trapping in thin-film silicon solar cells with improved self-assembled silver nanoparticles. *Nano Lett.* **2012**, *12*, 4070–4076. [CrossRef]
59. Ferry, V.E.; Verschuuren, M.A.; Lare, M.C.V.; Schropp, R.E.; Atwater, H.A.; Polman, A. Optimized Spatial Correlations for Broadband Light Trapping. *Nano Lett.* **2011**, *11*, 4239–4245. [CrossRef]
60. Zhu, J.; Hsu, C.M.; Yu, Z.; Fan, S.; Cui, Y. Nanodome solar cells with efficient light management and self-cleaning. *Nano Lett.* **2010**, *10*, 1979–1984. [CrossRef]
61. Herkommer, A.M.; Reichle, R.; Häfner, M.; Pruss, C. Design and simulation of diffractive optical components in fast optical imaging systems. *Opt. Des. Eng. IV* **2011**, *8167*, 816708.
62. Wood, A.P. Design of infrared hybrid refractive–diffractive lenses. *Appl. Opt.* **1992**, *31*, 2253. [CrossRef]
63. Feng, L.; Xiping, X.; Xiangyang, S. Design of infrared (IR) hybrid refractive/diffractive lenses for target detecting/tracking. *Acta Opt. Sin.* **2010**, *30*, 2084–2088. [CrossRef]
64. Škereň, M.; Svoboda, J.; Květoň, M.; Hopp, J.; Possolt, M.; Fiala, P. Diffractive elements for correction of chromatic aberrations of illumination systems. *EPJ Web Conf.* **2013**, *48*, 00025. [CrossRef]
65. Flores, A.; Wang, M.R.; Yang, J.J. Achromatic hybrid refractive-diffractive lens with extended depth of focus. *Appl. Opt.* **2004**, *43*, 5618. [CrossRef]
66. Wyrowski, F.; Bryngdahl, O. Digital holography as part of diffractive optics. *Rep. Prog. Phys.* **1991**, *54*, 1481. [CrossRef]
67. Mait, J.N. Understanding diffractive optic design in the scalar domain. *JOSA A* **1995**, *12*, 2145–2158. [CrossRef]
68. Bryngdahl, O.; Wyrowski, F. I Digital Holography–Computer-Generated Holograms. *Prog. Opt.* **1990**, *28*, 1–86.
69. Pillai, S.; Catchpole, K.R.; Trupke, T.; Green, M.A. Surface plasmon enhanced silicon solar cells. *J. Appl. Phys.* **2007**, *101*, 9. [CrossRef]
70. Catchpole, K.R.; Polman, A. Design principles for particle plasmon enhanced solar cells. *Appl. Phys. Lett.* **2008**, *93*, 10–13. [CrossRef]
71. Beck, F.J.; Polman, A.; Catchpole, K.R. Tunable light trapping for solar cells using localized surface plasmons. *J. Appl. Phys.* **2009**, *105*, 114310. [CrossRef]
72. Mokkapati, S.; Beck, F.J.; Polman, A.; Catchpole, K.R. Designing periodic arrays of metal nanoparticles for light-trapping applications in solar cells. *Appl. Phys. Lett.* **2009**, *95*, 053115. [CrossRef]
73. Yang, Y.; Pillai, S.; Mehrvarz, H.; Kampwerth, H.; Ho-Baillie, A.; Green, M.A. Enhanced light trapping for high efficiency crystalline solar cells by the application of rear surface plasmons. *Sol. Energy Mater. Sol. Cells* **2012**, *101*, 217–226. [CrossRef]
74. Deetjen, T.A.; Conger, J.P.; Leibowicz, B.D.; Webber, M.E. Review of climate action plans in 29 major U.S. cities: Comparing current policies to research recommendations. *Sustain. Cities Soc.* **2018**, *41*, 711–727. [CrossRef]
75. Boriskina, S.; Zheludev, N.I. (Eds.) *Singular and Chiral Nanoplasmonics*; CRC Press: New York, NY, USA, 2014.
76. Gangopadhyay, U.; Dutta, S.K.; Saha, H. *Texturization and Light Trapping in Silicon Solar Cells*; Nova Science Publishers Inc: New York, NY, USA, 2009; pp. 44–62.
77. Kane, D.; Micolich, A.; Rabeau, J. *Nanotechnology in Australia: Showcase of Early Career Research*; Pan Stanford Publishing Pte. Ltd.: Singapore, 2011; p. 463.
78. Clarke, J. *Energy Simulation in Building Design*; Routledge: London, UK, 2001. [CrossRef]
79. Fonash, S.J. *Introduction to Light Trapping in Solar Cell and Photo-Detector Devices*; Elsevier: Amsterdam, The Netherlands, 2015.
80. Tiwari, A.; Boukherroub, R.; Sharon, M. *Solar Cell Nanotechnology*; Wiley-Scrivener: Austin, TX, USA, 2013.

© 2019 by the authors. Licensee MDPI, Basel, Switzerland. This article is an open access article distributed under the terms and conditions of the Creative Commons Attribution (CC BY) license (http://creativecommons.org/licenses/by/4.0/).

Review

Encapsulation of Organic and Perovskite Solar Cells: A Review

Ashraf Uddin *, Mushfika Baishakhi Upama, Haimang Yi and Leiping Duan

School of Photovoltaic and Renewable Energy Engineering, University of New South Wales, Sydney 2052, Australia; m.upama@unsw.edu.au (M.B.U.); haimang.yi@student.unsw.edu.au (H.Y.); leiping.duan@unsw.edu.au (L.D.)
* Correspondence: a.uddin@unsw.edu.au

Received: 29 November 2018; Accepted: 21 January 2019; Published: 23 January 2019

Abstract: Photovoltaic is one of the promising renewable sources of power to meet the future challenge of energy need. Organic and perovskite thin film solar cells are an emerging cost-effective photovoltaic technology because of low-cost manufacturing processing and their light weight. The main barrier of commercial use of organic and perovskite solar cells is the poor stability of devices. Encapsulation of these photovoltaic devices is one of the best ways to address this stability issue and enhance the device lifetime by employing materials and structures that possess high barrier performance for oxygen and moisture. The aim of this review paper is to find different encapsulation materials and techniques for perovskite and organic solar cells according to the present understanding of reliability issues. It discusses the available encapsulate materials and their utility in limiting chemicals, such as water vapour and oxygen penetration. It also covers the mechanisms of mechanical degradation within the individual layers and solar cell as a whole, and possible obstacles to their application in both organic and perovskite solar cells. The contemporary understanding of these degradation mechanisms, their interplay, and their initiating factors (both internal and external) are also discussed.

Keywords: organic solar cells; perovskite solar cells; encapsulation; stability

---

1. Introduction

Organic photovoltaic (OPV) and Perovskite solar cell (PSC) are promising emerging photovoltaics thin film technology. Light harvester metal-halide perovskite materials, such as methyl-ammonium lead iodide ($CH_3NH_3PbI_3$), have exhibited small exciton binding energy, high optical cross-section, superior ambipolar charge transport, tuneable band gaps, and low-cost fabrication [1]. The $CH_3NH_3PbI_3$ PSCs and OPV devices can be solution-processed, which is suitable for roll-to-roll manufacturing processes for inexpensive largescale commercialization. The power conversion efficiency (PCE) of OPV devices have overpassed 14% for single junction and 17% for tandem devices to date with the development of low band-gap organic materials and device processing technology [2–5]. The achievement of the highest PCE of PSCs over 23.3% has shown promising future directions for using in large-scale production, together with traditional silicon solar cells [6]. Low-temperature (<150 °C) and solution-processed ZnO based electron transport layer (ETL) is one of the most promising materials for large scale roll-to-roll fabrication of perovskite and organic solar cells, owing to its almost identical electron affinity (4.2 eV) of TiO_2. Arafat et al. have already demonstrated aluminium (Al) doped ZnO (AZO) ETL of PSC with cell efficiency over 18% [7]. Currently, the poor stability of OPV and perovskite solar cells is a barrier for the commercialisation [8]. It is believed that oxygen and moisture are the external main reason for degradation of organic and PSCs, as shown in Figure 1 [9]. All the internal possible degradation mechanisms in PSCs can be controlled by careful interface engineering, such as a good choice of cathode and anode interlayer materials, ion-hybridizations in perovskite layer, etc.

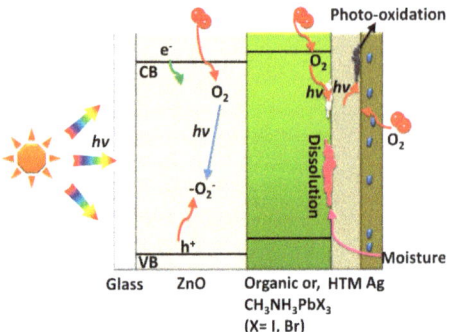

Figure 1. Schematic diagram of an organic/Perovskite solar cell (PSC) solar cell structure. The electron-hole pair recombination, moisture dissolution of perovskite material and photo-oxidation processes at the interface between hole transport materials (HTM) and metal electrode are shown. Adapted with permission from [9]. Copyright 2018 Royal Society of Chemistry.

Encapsulation of organic and perovskite solar cells can play an effective role in improving the stability of both devices. The encapsulation layer can act as a barrier layer by restricting the diffusion of oxygen and moisture through this encapsulation material, resulting in the protection of the cathode interface and the active layer from deterioration as shown in Figure 2. Encapsulation materials should have high barrier performance for oxygen and moisture. The encapsulation material layer structure is a critical factor to overcome these issues and enhance the device stability [10].

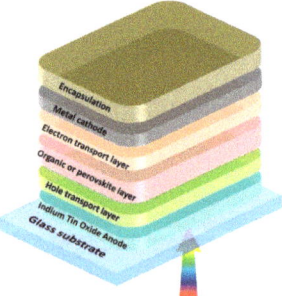

Figure 2. A schematic diagram of an organic or perovskite solar cells structure with an encapsulation layer.

The encapsulation material needs to possess good process ability, low water absorptivity, and permeability. Encapsulation materials should have relatively higher dielectric constant, light transmission, and resistance to ultraviolet (UV) degradation and thermal oxidation. They also require excellent chemical inertness, adhesion, and mechanical strength [11,12]. Oxygen transmission rate (OTR) and water vapour transmission rate (WVTR) are the steady state rates at which oxygen and water vapour gas can penetrate through a film that affects the encapsulation layer. A schematic diagram of organic and/or perovskite solar cells device is shown in Figure 2 with an encapsulation layer. This encapsulation layer material should have the following (Table 1) required properties to protect the device from the oxygen and water vapour effects. A list of specifications of requirements of encapsulation materials is listed in Table 1.

Table 1. Specifications and requirements for encapsulating materials for the protection of organic and perovskite devices from the oxygen and moisture. Reprinted with permission from [12]. Copyright 2016 John Wiley & Sons, Inc.

Characteristics	Specification of Requirement
WVTR	10^{-3}–10^{-6} g·m^{-2}·day^{-1}
OTR	10^{-3}–10^{-5} cm^3·m^{-2}·day^{-1}·atm^{-1}
Glass transition temperature (T_g)	<−40 °C (during the winter in deserts)
Total light transmission	>90% of incident light
Water absorption	<0.5 wt % (20 °C/100% RH)
Tensile modulus	<20.7 mPa (>3000 psi) at 25 °C
UV absorption degradation	None (>350 nm)
Hydrolysis	None (80 °C, 100% RH)

To determine the value of OTR, it is necessary to use a colorimetric sensor [13,14]. The OTR was calculated by measuring the amount of oxygen during a certain period at a constant rate that it passed through the cathode. By knowing the value of OTR and WVTR it is possible to determine how good is the encapsulated materials performance for the protection of perovskite and OPVs devices from degradation through their lifetime and performance. The effective WVTR can be determined by monitoring the temporal rate of change of the cathode (metal) electrical conductance. If it is assumed that the diffusivity of water vapour and oxygen obeys Fick's law (diffusivity is independent of concentration), then the WVTR or OTR can be described as [15]:

$$\text{WVTR}(t) = \frac{DC_s}{l}\left[1 + 2\sum_{n=1}^{n}(-1)^n e^{\frac{[-Dn^2\pi t]}{l^2}}\right] \quad (1)$$

where D is the diffusivity, C_s is the surface saturation concentrations, t is the time, and l is the film thickness. Normally, D increases with increasing temperature. WVTR and OTR is also a function of temperature.

The good performance encapsulation materials should have a value larger than 10^{-6} g·m^{-2}·day^{-1} for WVTR (Table 1) to protect the OPV and PSCs devices [16]. As an example, an OPV device with a structure ITO/ZnO/P3HT:PCBM/PEDOT:PSS/Al was encapsulated using ZnO and UV resin with a large WVTR value of 5.0×10^{-1} g·m^{-2}·day^{-1} [17]. In Ref. [18], the WVTR value was found as big as of 100 g·m^{-2}·day^{-1} for an encapsulated OPV (ITO/PEDOT:PSS/P3HT:PC$_{71}$BM/Ca/Al) with an epoxy resin. On the other hand, in Ref. [17] OPVs cells under the configuration ITO/DMD/Cs$_2$CO$_3$/P3HT:PCBM/MoO$_3$/Al were encapsulated using polyvinyl butyral (PVB), ethylene vinyl acetate (EVA), and thermoplastic poly-urethane (TPU) with a WVTR value of 60, 40, and 150 g·m^{-2}·day^{-1}, respectively, and with an OTR value in the range of 10^{-2}–10^2 cm^3·m^{-2}·day^{-1}. In this current work, the WVTR and OTR values were not calculated due to different experimental details.

The most common type of an encapsulation method refers to thin film layers encapsulated on top of OPV and perovskite devices using atomic layer deposition (ALD) [19,20]. ALD is particularly suitable for organic and flexible electronics. However, the ALD technique is expensive. Other methods are roll lamination systems encapsulating the OPV devices between two sheets uniting them with an adhesive [21], the other method is based on heat sealing, a process that basically consists of supplying thermal energy on outside of package to soften/melt the sealants [14] and using a glass substrate that is to be sealed with thermosetting epoxy, it could not be effectively applied to flexible devices [22].

Many works have been done for the advancement of good quality encapsulation technologies to stop the migration of environmental oxygen and moisture into device layers. The device lifetime and stability can be improved with high barrier performance encapsulation materials and structures for providing sufficient durability for commercial application. The encapsulation materials with high optical transmission and high dielectric constant need to possess good processing ability. Mechanical strength, good adhesion, and chemical inertness are also required for a suitable encapsulation material.

It is also expected to have low water-absorptivity and permeability and relatively high resistance to UV degradation and thermal oxidation [23]. Organic materials with a good combination of these properties are most commonly used as the encapsulate for the improvement of acceptable device durability [24]. When compared to their inorganic counterparts, the organic encapsulation materials have demonstrated notable advantages, such as the flexible synthesis of the organic molecules by varying their energy levels, molecular weight, and solubility [25]. Organic encapsulation materials are also expendable and they have lesser impact on the environment [26]. Hence, organic materials are the best suitable candidate for the encapsulation of flexible organic and perovskite devices.

2. Degradation Mechanism of OPV and Perovskite Solar Cells

The degradation of organic and perovskite devices can be divided into two mechanisms: intrinsic and extrinsic. Both of these mechanisms are related to mass-transport processes. The metal electrode/active layer interfaces play major roles in the degradation of both devices [27]. Under normal environmental conditions, the relatively high sensitivity of organic and perovskite materials towards oxygen and moisture decrease the device reliability and lifetime.

2.1. Degradation of OPV Devices

Intrinsic degradation: It is due to changes in the structures of the interfaces between layers of the stacking materials, owing to internal modification of the materials used. The key issue is stability, which is limiting the practical applications of OPV devices. The lifetime of OPV devices is now over many thousands of hours. These improvements of lifetime have been achieved with the application of device architecture optimizations and different interface materials, especially after the investigation of hole conductor layers incorporating carbon electrodes. This is promised stable, low cost, and easy device fabrication methods.

Extrinsic Stability: It is affected by the infiltration of air, e.g., oxygen and water. Such degradation from external factors can escalate by light irradiation. There are some organic materials and metal electrodes can also be degraded by oxygen and water. Oxygen or moisture can be trapped during the fabrication processes or they could diffuse into the cell structure during device lifetime. Due to the following factors, oxygen strongly influences the extrinsic stability in some OPV devices [28]: (i) fullerene molecules are hydrophobic and does not react with water; (ii) the exposure to oxygen in air has negative impact on the electron-transport properties of fullerene; and (iii) oxygen forms surface dipoles and increases the metal work-functions. The abovementioned properties yield in the deterioration of the performance of conventional OPVs; however, they might initially enhance the performance of inverted OPV [11]. The chemical degradation processes are the degradation of the metallic electrodes, degradation of the transparent electrode, intermediate hole extraction layers, and/or even the chosen method to synthesize the materials [29,30]. Although it is well-known that oxygen is typically a p-type dopant in semiconductors, the oxygen vacancies act as electron donors [31]. The electronic properties of semiconducting p-type layers might be improved briefly upon exposure to oxygen [32]. To overcome this issue, researchers have developed inverted OPV devices [33,34]. In a humid environment, the oxide layer can double in thickness, and hence block charge-tunnelling. Unlike the abovementioned positive/negative effects of oxygen, there is hardly any positive impact of water on OPV devices that has ever been suggested. Encapsulation delays the process of degradation, but the currently available materials used for encapsulation. The overall device degradation is not stopped, even by a complex encapsulation schemes, such as a sealed glass container or a high vacuum chamber are employed, because the processes involving water and oxygen are efficiently alleviated. The physical and chemical characteristics of the constituent materials are a complex phenomenon. Several processes may take place simultaneously in both physical and chemical characteristics [35].

2.2. Degradation of Perovskite Devices

There are several issues that are related to the degradation of PSCs, such as interface and device instability. These must be addressed to achieve good reproducibility with high conversion efficiencies and long lifetimes of PSCs. A comprehensive understanding of these issues in PSCs is required to achieve stability breakthroughs for practical commercial applications. For the PCE improvement, the stability of PSCs has to be improved for successful small- and large-scale applications.

The degradation of PSCs can occur as a result of intrinsic and extrinsic stability, such as thermal instability (intrinsic stability) and susceptibility of the perovskite material to ambient conditions (e.g., oxygen and humidity). Degradation can occur by other device structure components, such as the degradation of the ETL (TiO_2 or ZnO) at the interface upon light irradiation and the lack of stability of the hole-transport material [36,37]. The stability of perovskite thin films and PSCs has been widely studied [38–40], which includes the degradation of the perovskite material upon exposure to illumination, humidity, or increased temperature [41–44]. Different schemes have been investigated to enhance the stability of PSCs, such as replacing the mesoporous layer [45], modifying the composition or deposition process of perovskite material [46–49], using various charge-transport materials [50–55] and carbon-based electrodes [46], and interfacial layers and/or surface treatments [56–59]. In most of the researches on stability, the stability tests were reported after the devices being stored in the ambient or dark with measurements of the device performance at regular intervals [52,54,55,60]. The report on encouraging stability of the perovskite devices, such as 0.3% PCE drop in one month, often involves storing them under dark in a low-humidity desiccator [39].

However, it is now well established that PSCs are very susceptible to high humidity. The degradation is notably quicker under high humidity conditions [61,62], and it is much faster under continuous illumination than degradation involving storage in the dark [56,62]. Even the degradation under ambient illumination is significantly faster when compared to dark storage [63]. Since there is no standard testing protocol for reporting stability test results, it is strenuous to draw comparisons directly between test results from different publications. For example, Zhu et al. performed stability tests under constant illumination over a short time period in an N_2 glovebox to prevent the intrusion of humidity [62]. Thermal stability has been reported without exposure to illumination or humidity, or a combination of thermal stress and illumination [64]. In some studies, a white light-emitting diode (LED) was used for the illumination, not a solar simulator [50]. The stability can be over-estimated under illumination, owing to the absence of strong UV component in the emission spectrum. The degradation of the perovskite film is reportedly slower under illumination when a UV filter is used in order to eliminate the UV component from the simulated sunlight illumination [43]. Some works have applied several stress factors at the same time during the tests. The factors include constant illumination, ambient humidity of +50%, and increased temperature [65]. Moreover, the moderately humid environment has been used to obtain outdoor testing results [66].

Multiple stability tests have been performed on cells under low ambient humidity without encapsulation, typically under dark conditions [38,46–49], with very few exceptions [50,57,67]. After several hours, devices without encapsulation generally displayed severe degradation under continuous illumination whilst encapsulated devices showed a relatively longer lifetime [39]. In comparison to fully encapsulated devices with a protected edge, partially encapsulated devices showed a shorter lifespan [46]. This signifies the importance of complete encapsulation in order to enhance extend the lifetime of OPV and perovskite devices. However, the comprehensive comparison between different encapsulation techniques for PSCs is still rare [65]. A careful and complicated encapsulation system, developed by employing the combination of a desiccant material and UV epoxy resin, exhibited significantly improved stability. However, the device performance quickly decreased to <50% of the initial efficiency within 10 h under constant illumination and at an ambient humidity of 80% and cell temperature of 85 °C [65]. The absence of standardized testing conditions and studies regarding the effect of encapsulation on the device stability is hampering the advancements of stable organic and perovskite cells. Various encapsulation methods have been investigated for common perovskite devices that include planar TiO$_2$ on fluorine doped tin oxide (FTO) glass, methylammonium lead iodide (MAPbI$_3$) as the active layer, and tetrakis[N,N-di(4-methoxyphenyl) amino]-9,9'-spirobifluorene (spiro-OMeTAD) as the hole transport layer (HTL) [43]. The international summit on OPV stability (ISOS) was followed for the organic solar cell testing (ISOS-L-2 and ISOS-O-1) [68].

2.2.1. Thermal and Photo Stability

It is found that the photo-stability of perovskite materials is a serious problem for the application of PSCs. The perovskite material begins to decompose into PbI$_3$ at temperature 140 °C [69], and even at temperature 85 °C when it is heated in nitrogen for 24 h [41]. The slight heating of perovskite materials improved the performance of devices due to the formation of a small amount of PbI$_3$ which passivate the perovskite. The device performance may be recovered if it is stored in the dark for a short time. Poor photo-stability of perovskite materials could be raised from the local phase change under an elevated temperature when exposed to light. The dual ion hybridization by compositional material engineering at two sites (A and X) simultaneously can solve the perovskite materials stability. The existing ionic components within the perovskite crystal structure would be partially substituted by analogous ions of similar electronic properties, which would passivate the vulnerable sites present within the structure and thereby eliminating the degradation pathways efficiently from the inside-out. For example—the existing monovalent (A-site) organic cation would be partially substituted by metal (Caesium, Cs$^+$) ions. It has been experimentally proved that heat rather than moisture was the main cause of PSC degradation [70]. The key stability of PSC is to prevent the escape of volatile decomposition products from the perovskite solar cell materials. Polyisobutylene (PIB) encapsulation is one of the promising low cost and low application temperature packaging solutions for PSCs.

2.2.2. Ion Movement

Ion migration in the perovskite materials is another critical issue for the PSCs [71]. The ions (anion or cation) in the halide perovskite materials can move under bias voltage or thermal drift, causing device instability. A schematic diagram of different ion/defects movements in perovskite device structure is shown in Figure 3.

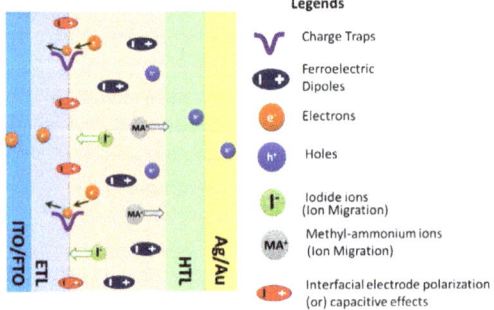

Figure 3. Schematic diagram shows the various ion/defects movements in perovskite device structure that could possibly lead to hysteresis phenomenon during current-voltage characteristic measurements. The exact nature of each ion-movement processes and its degree of influence on hysteresis varies with respect to the device structure and materials involved. Reprinted with permission from [71]. 2018 Solar Energy Materials and Solar Cells.

This defects/ion migration, such as iodine vacancies across the interface, can induce interfacial degradation. This defect migration affects device operational mechanisms and finally cause device failure during operation [9,72].

2.2.3. Electrode Degradation

The interfacial stability of PSCs is also critical for the overall device stability. Recently, it has been found that reactions between iodine-based decomposition products, such as HI from perovskite materials, and traditional metal electrode materials, such as Al or Ag, can lead to the formation of AlI$_3$ or AgI, respectively [55]. The formation of AlI$_3$ or AgI compound escalates perovskite decomposition when the material is exposed in an ambient environment (Figure 3) [55]. One way of enhancing the device stability is to insulate the perovskite material from the electrode (Ag/Al). The dispersal of halide ions in PSCs through the charge transport layers could exert a negative impact on the long-term stability of cells.

2.2.4. Charge Transport Layers Degradation

The reasons behind poor device stability of PSCs are due to either the perovskite materials or charge transport layer interfaces. Organic semiconductor layers in the PSC structure that are used as charge transport layers (ETL and HTL) are prone to both oxidization and water absorption that can reduce the device stability. For example, an n-type fullerene, PCBM was used as an efficient ETL in an inverted structure PSC [73]. The authors found that the PCBM went through changes in terms of chemical states or band structure in ambient air, which was a major contributing factor to the device degradation [73]. A p-type organic semiconductor, PEDOT:PSS, is generally used in the inverted structure as the HTL. PEDOT:PSS has high water absorptivity and acidic properties causing etching of the transparent conductive electrode, such as ITO. The hygroscopic and acidic nature of PEDOT:PSS accelerates device degradation [55,74]. Lithium (Li) doped Spiro-OMeTAD salt is also hygroscopic and can diffuse water into the light absorbing perovskite layer causing device failure [75]. Several investigations suggested that oxygen de-absorption from TiO$_2$ surface (degradation of TiO$_2$ layer), while light soaking leads to failure of devices [45]. Multiple techniques have been employed for the improvement of the device stability [7,74]. Idigoras et al. [76] deposited thin polymer layer by the remote plasma vacuum deposition of adamantine powder for the encapsulation of perovskite solar cells. They observed that the deterioration of device performance was significantly delayed because of the encapsulation layer.

Bella et al. [77] investigated the stability of PSCs by encapsulation of photochemical resistance as well as moisture tolerance, a fluoro-polymeric light-curable coating on the back contact side. The use of the fluoro-polymeric layer on top of the PSC has the prospect of preventing the oxygen and moisture migration between the top back contact and the perovskite layer due to the hydrophobicity of the coating. The fluoro-polymeric UV-coating possesses a cross-linked nature that can lower the free volume when compared to the traditional non–cross-linked, polymeric systems. Hence, it is expected to improve the long-term stability of PSCs [78]. Figure 4 displays the test results of device stability in terms of PCEs under various atmospheric conditions and photochemical external stresses during 180-days (4320 h). Moreover, those encapsulated devices were tested in outdoor conditions for a time of >3 months (i.e., 2160 h). Successful tolerance toward heavy rain, soil, and dust were observed on the external surface made of glass. In real outdoor conditions, the fluorinated luminescent down shifting layer with low-surface energy helps to clean the front electrode easily.

Uncoated and front-coated devices were studied for an overall period of six months in the aging test. The devices were stored in a glove box filled with Ar for the first three months. The devices were continuously illuminated for 8 h every day with a UV optical fibre with light intensity 5 mW·cm^{-2}. 5% contribution from the UV light source (280–400 nm) helped to simulate the solar spectral irradiance on Earth (AM1.5G, 1000 W·m^{-2}). In the forced UV aging test (Figure 4), the uncoated devices (black curve) had a reduction of 30% in their initial efficiency after one week of UV exposure and stopped working after one month. All of the five front-coated cells (red curve) retained 98% of their initial PCE, even after three months and demonstrated excellent stability under the same conditions.

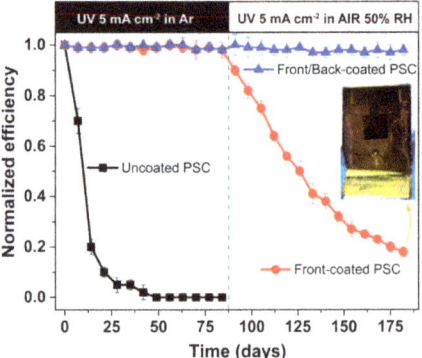

Figure 4. The aging test results on the three series of PSCs: (i) uncoated, (ii) front-coated with a luminescent fluorinated coating, and (iii) front/back-coated (front coating with the luminescent fluorophore and back contact coating with a moisture resistant fluoro-polymeric layer). The cells were stored in an Ar environment during the first three months of the test period. They were kept in the air with RH = 50% for the next three months. During the whole testing period, the PSCs were exposed to continuous UV illumination. The device power conversion efficiency (PCE) was recorded once a week. The inset shows a photograph of a front-coated solar cell at the end of the test. Reprinted with permission from [77]. Copyright 2018 American Association for the Advancement of Science.

By simple visual inspection, the effects of degradation were easily detectable. The progressive yellowing was observed in the mixed-perovskite layer upon 50% RH exposure. The back-contact side was coated with a fluoropolymeric light-curable coating in the front/back-coated devices for stabilizing the devices with enhanced photochemical resistance and moisture tolerance. A similar method was used in the luminescent downshifting experiment for back-coated PSCs. It is hypothesized that the use of highly hydrophobic fluoropolymeric layer on the PSC top can prevent water permeation from the top back contact to the perovskite layer. In addition, the fluoro-polymeric UV-coating possesses a cross-linked nature, leading to lower free volume when compared to the traditional non–cross-linked, polymeric systems. Hence, the coating is expected to improve the long-term stability of PSCs [79]. The aging test results of the front/back-coated PSCs (blue curve), i.e., devices with a luminescent coating on front and a moisture-resistant coating on the back contact, is shown in Figure 4. Even under the combined effects of photochemical and environmental stresses, all the five devices retained exceptional stability of 98%.

The front/back-coated PSCs showed longer lifetime stability retaining most of their initial efficiency the ageing test due to the following reasons:

- securing the perovskite material from UV irradiation and converting it into visible photons;
- shielding devices from oxygen and moisture, hence blocking the hydrolytic behaviour of the perovskite material;
- maintaining clean front electrode clean by the self-cleaning characteristics of this fluorinated polymer. Similar results were also observed for outdoor tests performed.

The non-homogeneous deposition of the coating layers on the front/back side of the device can cause slow hydrolysis of the active perovskite layer. To test the homogeneity of the deposited coating layer, Bella et al. [77] dipped the devices with front/back coatings into the water for about a day. No change in photovoltaic performance was observed in the front/back-coated devices, even after one day of immersion in water.

3. Encapsulation Requirements

To ensure satisfactory encapsulation (Table 1), the permeation of WVTR and OTR needs to be two-three orders of magnitude lower than the bare substrate. Silicon-based dielectric films, deposited by PECVD, have reportedly demonstrated excellent transparent, single-layer barrier performance [80]. However, on polymer substrates it is very difficult to obtain permeation levels that are 1000 times lower by a single layer barrier. The increase in single-layer barrier thickness results in an OTR value that is on-zero asymptotic and is governed by the defects in the barrier layer. Such defects are typically generated from the intrinsic or extrinsic roughness of surface [81]. Hence, the performance of single-layer, encapsulated gas barriers is controlled by the nano-meter size structural defects. In theoretical calculations, the total permeation rate is much higher due to a large number of small pin-holes than that from few but larger defects in the same total pin-hole area [82]. The lateral diffusion of gas is more crucial when the defect diameters are smaller than the substrate thickness. The multilayered deposition of inorganic materials can slightly enhance the performance of the encapsulation barrier. The deposition of organic–inorganic multilayers films to obtain increased barrier performance is the most common practice [83]. However, there are several challenges in the way of utilization and development of thin-film encapsulation. To face these challenges, a variety of materials have been developed to improve the device lifetimes, such as titanium oxide (TiO_x). Using thin films as barriers for device encapsulation is not sufficient for the protection of organic and perovskite devices. The use of glass plates as encapsulation is the simplest and best example for the protection of oxygen and moister. Glass encapsulation can supply the required oxygen and moisture protection in the organic and perovskite solar cells for commercial application. However, glass encapsulation cannot apply on inflexible devices. It has also been shown that a thin layer of SiO_xC_y acts as a barrier for oxygen and moisture. Unfortunately, the PECVD process is required for the deposition of this encapsulation film, which is expensive and uses complicated vacuum systems. This is also not suitable to deposit on a flexible substrate as a barrier of oxygen and moister transmission. Although there are other encapsulating materials, they cannot fulfil the barrier requirements for organic and perovskite solar cells. In this review paper, we have presented different encapsulation materials used in various approaches and integrated with organic and perovskite devices. In Table 2, we have summarized some of the polymer materials with their WVTR values for the organic and perovskite solar cells. Recently, using the RF plasma polymerization technique linalyl acetate [84,85] and terpinene-4-ol thin films has been deposited for the encapsulation of devices. On the basis of surface and optical properties, these polymer based thin films can be potential encapsulating layers for organic and perovskite solar cells [86,87].

Materials for Encapsulation

The encapsulation materials should have high a dielectric breakdown that matches the refractive index with other layers and high volume resistivity. Materials should be low cost, dimensionally stability, and easy to deposit. WVTR and OTR represent the steady state rates at which water vapour and oxygen gas, respectively, penetrate through the encapsulating film that affects the encapsulation layer. Glass transition temperature (T_g) for organic encapsulation material is another important property, which is dependent on the chain flexibility. The polymer mechanically varies from being rigid and brittle and becomes tough and leathery. The maximum exposure temperatures on encapsulation material and the effect on the mechanical behaviour of the material should be known. The light transmission through the encapsulation materials is also important to measure to understand how it will affect the device performance. The encapsulation material requirements are also to define the UV absorption degradation, hydrolysis, and some other aspects. A list of suitable polymer encapsulation materials and their specifications for the roll-to-roll fabrication of devices are given below.

Table 2. List of probable polymer encapsulation materials for organic photovoltaic (OPV) and Perovskite devices.

Materials	Encapsulation Type	Water Vapour Transmission Rate (WVTR) ($g \cdot m^{-2} \cdot day^{-1}$)	Comments	References
Ethylene vinyl acetate (EVA)	Single layer encapsulation	40	Light transmission of 91%. Suitable for resisting weather and long-term reliability under light exposure. It is suitable for encapsulation of organic and perovskite solar modules.	[88,89]
Europium doped EVA: Eu^{3+}	Single layer encapsulation	40	Absorption bang gap is 310 nm (4 eV). Suitable for PV module encapsulation. Eu^{3+} doped EVA layers can induce photon down-shifting with wavelengths <460 nm.	[90]
Ethylene methyl acrylate (EMA)	Single layer encapsulation	Not mentioned	EMA is suitable for chemical resistance, thermally stability, adherence to different substrates and excellent mechanical behaviour at low temperature. It is suitable for encapsulation of perovskite and organic devices.	[23,91]
Polyvinyl butyral (PVB)	Single layer encapsulation	60	PVB has high optical transparency, good heat resistance, good adherence to solar cells, glass, and other plastics, increased bond durability, and compatible with module components. PVB is already used as encapsulation layer for thin film solar cells.	[14]
Thermoplastic polyurethane (TPU)	Single layer encapsulation	150	TPU film is better than EVA film for encapsulation since it is flexible to bond with relatively hard materials. These films can be processed in normal atmospheric pressure without cross-linking and emissions.	[92,93]
UV-cured epoxy	Single layer encapsulation	16	Epoxy film is good for encapsulation. It is optically transparent, thermally conductive, weather resistance, high temperature resistance, good adhesive properties on glass and plastic.	[70]
Polyisobutylene (PIB)	Single layer encapsulation	0.001–0.0001	PIB is a synthetic rubber. It can encapsulate organic and perovskite solar cells.	[70]
Cyclized perfluoro-polymer (CytopTM)	Single layer encapsulation	Not mentioned	Conventional thin-film deposition techniques e.g., spin coating can be used to deposit this polymer. It is transparent and amorphous. This material is good for weather resistant, good for oxygen/water-vapor shielding for testing organic device lifetime.	[94]
Organic–inorganic hybrid materials ORMOCERs (ORM)	Single layer encapsulation	0.01	It is organic and inorganic components modified ceramics or ORM. It has good chemical resistance, highly transparent, anti-soiling, diffusion- inhibition. OTR is <0.01 $cm^3 \ m^{-2} \ day^{-1}$. These properties are necessary for the encapsulation of organic and perovskite solar cells.	[95,96]
ORMOSIL aero-gel thin film	Single layer encapsulation	Not mentioned	It is mechanically and thermally stable and highly transparent. ORMOSIL materials are flexibility and stability at atmospheric conditions. ORMOSIL has variable organic group that can modify the chemical and physical surface properties.	[97,98]
Other organic materials	Single layer encapsulation	Not mentioned	Encapsulation tested of 10 polymers were poly(vinyl methyl ketone) (PVMK), poly(methyl methacrylate) (PMMA), poly(vinylidene chloride-co-vinyl chloride) (PVDC-co-PVC), poly (vinylidene fluoride)(PVDF), polyacrylonitride (PAN), poly(vinylalcohol) (PVA), poly(vinylphenol) (PVP), poly(methyl vinyl ether) (PMVE), polystyrene (PS), and poly(vinyl chloride) (PVC). An encapsulation approach with these 10 low-polarity polymer is demonstrated to block water/moisture and prevent encapsulation-induced degradation.	[99,100]
Luminescent downshifting fluoro-polymeric coating such as PTFE (polytetrafluoroethylene), PFA (perfluoroalkoxy), FEP (fluorinated ethylene-propylene), etc.	Single layer encapsulation	Not mentioned	It is very good as an encapsulating material for organic and perovskite solar cells to improve the device stability for the out-door application. Fluoropolymers are excellent for chemical and thermal resistance. Their surfaces are non-reactive with all chemicals and solvents.	[101]

4. Discussion

Thin-film encapsulation is a vital technology that is required for the application and commercialization of perovskite and organic solar cells. An effective encapsulation is crucial to prevent the permeation of moister and oxygen to achieve the desired reliability and device lifetime. More progress is needed to develop encapsulating materials for devices with specific requirements. The processing conditions of EVA exposure time and damp heat affect its adhesion strength. Transport of moister and oxygen through local pinholes become an issue for achieving ultrahigh barriers encapsulation that is hard to be avoided. New encapsulation materials need to be developed to fulfil the encapsulation requirements and achieve ultrahigh encapsulation barriers. Furthermore, the encapsulation processing temperature of thin film should be within the temperature range that is suitable for the organic polymer substrate materials. Usually, low-temperature PECVD processed inorganic encapsulation thin films are not suitable for the encapsulation barrier, since these films suffer from intrinsic defects. Water vapour and oxygen can easily diffuse through these defects. For high performance encapsulation barriers, these intrinsic defects should be reduced or passivated. Inorganic and organic alternating multilayer films can be one solution to reduce and avoid intrinsic pinholes. The defects in the inorganic layers can be passivated by an organic film and do not channel continuously through the multilayer structure. A steady state permeation rate of moister and oxygen can be maintained by the multilayer encapsulation barrier structures. A transient rate of encapsulation barrier can be maintained over a specific time-period that might exceed the desired lifetime of encapsulated devices. Typically, the permeation rate in the transient region is lower than the permeation rate in the steady state. It is suggested that the barrier performance characterization from the initial transient period will provide an underestimation of the total permeation rate for long-term applications. As a result, the barrier performance in multilayer structures should be characterized separately, such as the steady-state and transient permeation rates. This should be used for the calculation of lag time to avoid overestimation of expected device lifetimes or barrier performance during both storage and application. Moreover, the deposition of high quality inorganic encapsulation layer with vacuum deposition processes is expensive and low through put.

Since there are alternative deposition processes using different deposition chambers for organic and inorganic layers, multilayer barrier structures are expensive. If inorganic layers can be deposited from the solution precursors, it can be a promising solution to this problem, making it possible to deposit inorganic and organic multilayer barriers employing similar deposition techniques. Hence, totally solution process should be developed for low cost, high barrier materials for perovskite and OPV encapsulation. Optimization of the individual film dimension can be useful multilayer encapsulation structures. Before adapting these technologies, the quality of films should also be improved. The reliability and continuous yield of encapsulation processes must be explored to reduce the processing steps and time. Since perovskite and organic solar cells are very susceptible to moisture, it is crucial to use a low-diffusivity edge seal material.

Graphene is a carbon-based one-dimensional material with excellent electronic and mechanical properties offering limitless opportunities in device engineering. In line with this, graphene can be used as a bi-functional electrode that will serve as a highly conductive charge collection electrode as well as an encapsulant for the underlying organic or perovskite layers. This carbon-graphene based encapsulation technology is sought to be a viable approach for large scale commercialization and deployment of organic and perovskite solar cells technology.

5. Conclusions

Effective thin film encapsulation is crucial to prevent the permeation of water vapour and oxygen for achieving the stability and desired life times of organic and perovskite solar cells. The problem of achieving a thin layer encapsulation barrier is transport-dependent permeation through localized pin-holes that is strenuous to control. New materials and technology are required to satisfy the requirements of encapsulation to prevent the permeation of moisture and oxygen for the reliability and stability of devices. The encapsulation processing temperature of thin film should be within the temperature range suitable for the organic and polymer substrate materials. Multilayer polymer films of encapsulation can be a solution to block the pinholes in the films to stop the permeation of water vapour and oxygen. However, the barrier performance of multilayer structures encapsulation consists of a steady state permeation rate as well as a transient rate over a specified time. These rates of encapsulation barrier can exceed the desired lifetime of encapsulated devices. The barrier performance in multilayer structures should be characterized separately by the steady-state and transient permeation rates over a specified time-period. Multilayer thin film encapsulation structures deposited under vacuum will be exorbitant, owing to alternative deposition of layers in separate deposition chambers. Solution process deposition can be a promising alternative for low-cost barrier materials for thin film multilayers encapsulation. Subsequently, multilayers can be deposited with a similar deposition process to obtain low-cost barrier materials. Film quality should be improved before adopting the multilayer encapsulation technologies. Encapsulation layers with high quality are significant to improve the device lifetimes.

Funding: This research received no external funding.

Acknowledgments: We acknowledge the financial support from Future Solar Technologies Pty. Ltd. for this work. We also acknowledge the endless support from the staff of photovoltaic and renewable energy engineering school (SPREE), UNSW, Sydney.

Conflicts of Interest: The authors declare no conflict of interest. The funders had no role in the design of the study; in the collection, analyses, or interpretation of data; in the writing of the manuscript, and in the decision to publish the results.

References

1. Liang, P.-W.; Liao, C.-Y.; Chueh, C.-C.; Zuo, F.; Williams, S.T.; Xin, X.-K.; Lin, J.; Jen, A.K.-Y. Additive Enhanced crystallization of solution-processed perovskite for highly efficient planar-heterojunction solar cells. *Adv. Mater.* **2014**, *26*, 3748–3754. [CrossRef] [PubMed]
2. Li, H.; Xiao, Z.; Ding, L.; Wang, J. Thermostable single-junction organic solar cells with a power conversion efficiency of 14.62%. *Sci. Bull.* **2018**, *63*, 340–342. [CrossRef]
3. Meng, L.; Zhang, Y.; Wan, X.; Li, C.; Zhang, X.; Wang, Y.; Ke, X.; Xiao, Z.; Ding, L.; Xia, R.; et al. Organic and solution-processed tandem solar cells with 17.3% efficiency. *Science* **2018**, *361*, 1094–1098. [CrossRef] [PubMed]
4. Xiao, Z.; Jia, X.; Ding, L. Ternary organic solar cells offer 14% power conversion efficiency. *Sci. Bull.* **2017**, *62*, 1562–1564. [CrossRef]
5. Xu, C.; Wright, M.; Ping, D.; Yi, H.; Zhang, X.; Mahmud, M.A.; Sun, K.; Upama, M.B.; Haque, F.; Uddin, A. Ternary blend organic solar cells with a non-fullerene acceptor as a third component to synergistically improve the efficiency. *Org. Electron.* **2018**, *62*, 261–268. [CrossRef]
6. Green, M.A.; Hishikawa, Y.; Dunlop, E.D.; Levi, D.H.; Hohl-Ebinger, J.; Ho-Baillie, A.W. Solar cell efficiency tables (version 51). *Prog. Photovolt. Res. Appl.* **2018**, *26*, 3–12. [CrossRef]
7. Mahmud, M.A.; Elumalai, N.K.; Pal, B.; Jose, R.; Upama, M.B.; Wang, D.; Goncales, V.R.; Xu, C.; Haque, F.; Uddin, A. Electrospun 3D composite nano-flowers for high performance triple-cation perovskite solar cells. *Electrochim. Acta* **2018**, *289*, 459–473. [CrossRef]
8. Wang, D.; Wright, M.; Elumalai, N.K.; Uddin, A. Stability of perovskite solar cells. *Sol. Energy Mater. Sol. Cells* **2016**, *147*, 255–275. [CrossRef]
9. Berhe, T.A.; Su, W.-N.; Chen, C.-H.; Pan, C.-J.; Cheng, J.-H.; Chen, H.-M.; Tsai, M.-C.; Chen, L.-Y.; Dubale, A.A.; Hwang, B.-J. Organometal halide perovskite solar cells: Degradation and stability. *Energy Environ. Sci.* **2016**, *9*, 323–356. [CrossRef]
10. Li, F.; Liu, M. Recent efficient strategies for improving the moisture stability of perovskite solar cells. *J. Mater. Chem. A* **2017**, *5*, 15447–15459. [CrossRef]
11. Ahmad, J.; Bazaka, K.; Anderson, L.J.; White, R.D.; Jacob, M.V. Materials and methods for encapsulation of OPV: A review. *Renew. Sustain. Energy Rev.* **2013**, *27*, 104–117. [CrossRef]

12. Kang, H.; Kim, G.; Kim, J.; Kwon, S.; Kim, H.; Lee, K. Bulk-heterojunction organic solar cells: Five core technologies for their commercialization. *Adv. Mater.* **2016**, *28*, 7821–7861. [CrossRef]
13. Cros, S.; De Bettignies, R.; Berson, S.; Bailly, S.; Maisse, P.; Lemaitre, N.; Guillerez, S. Definition of encapsulation barrier requirements: A method applied to organic solar cells. *Sol. Energy Mater. Sol. Cells* **2011**, *95*, S65–S69. [CrossRef]
14. Kim, N.; Potscavage , W.J., Jr.; Sundaramoothi, A.; Henderson, C.; Kippelen, B.; Graham, S. A correlation study between barrier film performance and shelf lifetime of encapsulated organic solar cells. *Sol. Energy Mater. Sol. Cells* **2012**, *101*, 140–146. [CrossRef]
15. Crank, J. *The Mathematics of Diffusion*; Oxford University Press: Oxford, UK, 1979.
16. Kippelen, B.; Brédas, J.-L. Organic photovoltaics. *Energy Environ. Sci.* **2009**, *2*, 251–261. [CrossRef]
17. Hauch, J.A.; Schilinsky, P.; Choulis, S.A.; Rajoelson, S.; Brabec, C.J. The impact of water vapor transmission rate on the lifetime of flexible polymer solar cells. *Appl. Phys. Lett.* **2008**, *93*, 103306. [CrossRef]
18. Lee, H.-J.; Kim, H.-P.; Kim, H.-M.; Youn, J.-H.; Nam, D.-H.; Lee, Y.-G.; Lee, J.-G.; bin Mohd Yusoff, A.R.; Jang, J. Solution processed encapsulation for organic photovoltaics. *Sol. Energy Mater. Sol. Cells* **2013**, *111*, 97–101. [CrossRef]
19. Kim, Y.; Kim, H.; Graham, S.; Dyer, A.; Reynolds, J.R. Durable polyisobutylene edge sealants for organic electronics and electrochemical devices. *Sol. Energy Mater. Sol. Cells* **2012**, *100*, 120–125. [CrossRef]
20. Zardetto, V.; Williams, B.; Perrotta, A.; Di Giacomo, F.; Verheijen, M.; Andriessen, R.; Kessels, W.; Creatore, M. Atomic layer deposition for perovskite solar cells: Research status, opportunities and challenges. *Sustain. Energy Fuels* **2017**, *1*, 30–55. [CrossRef]
21. Tanenbaum, D.M.; Dam, H.F.; Rösch, R.; Jørgensen, M.; Hoppe, H.; Krebs, F.C. Edge sealing for low cost stability enhancement of roll-to-roll processed flexible polymer solar cell modules. *Sol. Energy Mater. Sol. Cells* **2012**, *97*, 157–163. [CrossRef]
22. Elkington, D.; Cooling, N.; Zhou, X.; Belcher, W.; Dastoor, P. Single-step annealing and encapsulation for organic photovoltaics using an exothermically-setting encapsulant material. *Sol. Energy Mater. Sol. Cells* **2014**, *124*, 75–78. [CrossRef]
23. Cuddihy, E.; Coulbert, C.; Gupta, A.; Liang, R. *Electricity from Photovoltaic Solar Cells: Flat-Plate Solar Array Project Final Report*; Volume VII: Module Encapsulation; JPL Publication: Pasadena, CA, USA, 1986.
24. Kim, N. *Fabrication and Characterization of Thin-Film Encapsulation for Organic Electronics*; Georgia Institute of Technology: Atlanta, GA, USA, 2009.
25. Spanggaard, H.; Krebs, F.C. A brief history of the development of organic and polymeric photovoltaics. *Sol. Energy Mater. Sol. Cells* **2004**, *83*, 125–146. [CrossRef]
26. Shaheen, S.E.; Ginley, D.S.; Jabbour, G.E. Organic-based photovoltaics: Toward low-cost power generation. *MRS Bull.* **2005**, *30*, 10–19. [CrossRef]
27. Nguyen, T.-P.; Renaud, C.; Reisdorffer, F.; Wang, L.-J. Degradation of phenyl C61 butyric acid methyl ester: Poly (3-hexylthiophene) organic photovoltaic cells and structure changes as determined by defect investigations. *J. Photonics Energy* **2012**, *2*, 021013. [CrossRef]
28. Reese, M.O.; Morfa, A.J.; White, M.S.; Kopidakis, N.; Shaheen, S.E.; Rumbles, G.; Ginley, D.S. Pathways for the degradation of organic photovoltaic P3HT:PCBM based devices. *Sol. Energy Mater. Sol. Cells* **2008**, *92*, 746–752. [CrossRef]
29. Dennler, G.; Lungenschmied, C.; Neugebauer, H.; Sariciftci, N.S.; Latreche, M.; Czeremuszkin, G.; Wertheimer, M.R. A new encapsulation solution for flexible organic solar cells. *Thin Solid Films* **2006**, *511*, 349–353. [CrossRef]
30. Ray, B.; Alam, M.A. A compact physical model for morphology induced intrinsic degradation of organic bulk heterojunction solar cell. *Appl. Phys. Lett.* **2011**, *99*, 140. [CrossRef]
31. Karlsson, K.; Troncale, V.; Oberli, D.; Malko, A.; Pelucchi, E.; Rudra, A.; Kapon, E. Optical polarization anisotropy and hole states in pyramidal quantum dots. *Appl. Phys. Lett.* **2006**, *89*, 251113. [CrossRef]
32. Cao, H.; He, W.; Mao, Y.; Lin, X.; Ishikawa, K.; Dickerson, J.H.; Hess, W.P. Recent progress in degradation and stabilization of organic solar cells. *J. Power Sources* **2014**, *264*, 168–183. [CrossRef]
33. Lin, R.; Miwa, M.; Wright, M.; Uddin, A. Optimisation of the sol-gel derived ZnO buffer layer for inverted structure bulk heterojunction organic solar cells using a low band gap polymer. *Thin Solid Films* **2014**, *566*, 99–107. [CrossRef]

34. Lin, R.; Wright, M.; Chan, K.H.; Puthen-Veettil, B.; Sheng, R.; Wen, X.; Uddin, A. Performance improvement of low bandgap polymer bulk heterojunction solar cells by incorporating P3HT. *Org. Electron.* **2014**, *15*, 2837–2846. [CrossRef]
35. Grossiord, N.; Kroon, J.M.; Andriessen, R.; Blom, P.W. Degradation mechanisms in organic photovoltaic devices. *Org. Electron.* **2012**, *13*, 432–456. [CrossRef]
36. Tiep, N.H.; Ku, Z.; Fan, H.J. Recent advances in improving the stability of perovskite solar cells. *Adv. Energy Mater.* **2016**, *6*, 1501420. [CrossRef]
37. Ye, M.; Hong, X.; Zhang, F.; Liu, X. Recent advancements in perovskite solar cells: Flexibility, stability and large scale. *J. Mater. Chem. A* **2016**, *4*, 6755–6771. [CrossRef]
38. Bi, D.; Gao, P.; Scopelliti, R.; Oveisi, E.; Luo, J.; Grätzel, M.; Hagfeldt, A.; Nazeeruddin, M.K. High-performance perovskite solar cells with enhanced environmental stability based on amphiphile-modified $CH_3NH_3PbI_3$. *Adv. Mater.* **2016**, *28*, 2910–2915. [CrossRef] [PubMed]
39. Bi, D.; Tress, W.; Dar, M.I.; Gao, P.; Luo, J.; Renevier, C.; Schenk, K.; Abate, A.; Giordano, F.; Baena, J.-P.C. Efficient luminescent solar cells based on tailored mixed-cation perovskites. *Sci. Adv.* **2016**, *2*, e1501170. [CrossRef] [PubMed]
40. Shi, Z.; Jayatissa, A. Perovskites-based solar cells: A review of recent progress, materials and processing methods. *Materials* **2018**, *11*, 729. [CrossRef] [PubMed]
41. Conings, B.; Drijkoningen, J.; Gauquelin, N.; Babayigit, A.; D'Haen, J.; D'Olieslaeger, L.; Ethirajan, A.; Verbeeck, J.; Manca, J.; Mosconi, E. Intrinsic thermal instability of methylammonium lead trihalide perovskite. *Adv. Energy Mater.* **2015**, *5*, 1500477. [CrossRef]
42. Li, Y.; Xu, X.; Wang, C.; Wang, C.; Xie, F.; Yang, J.; Gao, Y. Degradation by exposure of coevaporated $CH_3NH_3PbI_3$ thin films. *J. Phys. Chem. C* **2015**, *119*, 23996–24002. [CrossRef]
43. Liu, F.; Dong, Q.; Wong, M.K.; Djurišić, A.B.; Ng, A.; Ren, Z.; Shen, Q.; Surya, C.; Chan, W.K.; Wang, J. Is excess PbI_2 beneficial for perovskite solar cell performance? *Adv. Energy Mater.* **2016**, *6*, 1502206. [CrossRef]
44. Niu, G.; Li, W.; Meng, F.; Wang, L.; Dong, H.; Qiu, Y. Study on the stability of $CH_3NH_3PbI_3$ films and the effect of post-modification by aluminum oxide in all-solid-state hybrid solar cells. *J. Mater. Chem. A* **2014**, *2*, 705–710. [CrossRef]
45. Leijtens, T.; Eperon, G.E.; Pathak, S.; Abate, A.; Lee, M.M.; Snaith, H.J. Overcoming ultraviolet light instability of sensitized TiO_2 with meso-superstructured organometal tri-halide perovskite solar cells. *Nat. Commun.* **2013**, *4*, 2885. [CrossRef] [PubMed]
46. Li, B.; Li, Y.; Zheng, C.; Gao, D.; Huang, W. Advancements in the stability of perovskite solar cells: Degradation mechanisms and improvement approaches. *RSC Adv.* **2016**, *6*, 38079–38091. [CrossRef]
47. Mahmud, M.A.; Elumalai, N.K.; Upama, M.B.; Wang, D.; Wright, M.; Chan, K.H.; Xu, C.; Haque, F.; Uddin, A. Single vs mixed organic cation for low temperature processed perovskite solar cells. *Electrochim. Acta* **2016**, *222*, 1510–1521. [CrossRef]
48. Xia, X.; Wu, W.; Li, H.; Zheng, B.; Xue, Y.; Xu, J.; Zhang, D.; Gao, C.; Liu, X. Spray reaction prepared $FA_{1−x}Cs_xPbI_3$ solid solution as a light harvester for perovskite solar cells with improved humidity stability. *RSC Adv.* **2016**, *6*, 14792–14798. [CrossRef]
49. Zhu, W.; Bao, C.; Li, F.; Yu, T.; Gao, H.; Yi, Y.; Yang, J.; Fu, G.; Zhou, X.; Zou, Z. A halide exchange engineering for $CH_3NH_3PbI_{3−x}Br_x$ perovskite solar cells with high performance and stability. *Nano Energy* **2016**, *19*, 17–26. [CrossRef]
50. Agresti, A.; Pescetelli, S.; Cinà, L.; Konios, D.; Kakavelakis, G.; Kymakis, E.; Carlo, A.D. Efficiency and stability enhancement in perovskite solar cells by inserting lithium—Neutralized graphene oxide as electron transporting layer. *Adv. Funct. Mater.* **2016**, *26*, 2686–2694. [CrossRef]
51. Arafat Mahmud, M.; Kumar Elumalai, N.; Baishakhi Upama, M.; Wang, D.; Haque, F.; Wright, M.; Howe Chan, K.; Xu, C.; Uddin, A. Enhanced stability of low temperature processed perovskite solar cells via augmented polaronic intensity of hole transporting layer. *Phys. Status Solidi (RRL)–Rapid Res. Lett.* **2016**, *10*, 882–889. [CrossRef]
52. Lee, D.-Y.; Na, S.-I.; Kim, S.-S. Graphene oxide/PEDOT:PSS composite hole transport layer for efficient and stable planar heterojunction perovskite solar cells. *Nanoscale* **2016**, *8*, 1513–1522. [CrossRef]
53. Wang, Y.; Yuan, Z.; Shi, G.; Li, Y.; Li, Q.; Hui, F.; Sun, B.; Jiang, Z.; Liao, L. Dopant-free spiro-triphenylamine/fluorene as hole—Transporting material for perovskite solar cells with enhanced efficiency and stability. *Adv. Funct. Mater.* **2016**, *26*, 1375–1381. [CrossRef]

54. Xu, J.; Voznyy, O.; Comin, R.; Gong, X.; Walters, G.; Liu, M.; Kanjanaboos, P.; Lan, X.; Sargent, E.H. Crosslinked remote-doped hole-extracting contacts enhance stability under accelerated lifetime testing in perovskite solar cells. *Adv. Mater.* **2016**, *28*, 2807–2815. [CrossRef] [PubMed]
55. You, J.; Meng, L.; Song, T.-B.; Guo, T.-F.; Yang, Y.M.; Chang, W.-H.; Hong, Z.; Chen, H.; Zhou, H.; Chen, Q. Improved air stability of perovskite solar cells via solution-processed metal oxide transport layers. *Nat. Nanotechnol.* **2016**, *11*, 75. [CrossRef] [PubMed]
56. Chang, C.-Y.; Chang, Y.-C.; Huang, W.-K.; Liao, W.-C.; Wang, H.; Yeh, C.; Tsai, B.-C.; Huang, Y.-C.; Tsao, C.-S. Achieving high efficiency and improved stability in large-area ITO-free perovskite solar cells with thiol-functionalized self-assembled monolayers. *J. Mater. Chem. A* **2016**, *4*, 7903–7913. [CrossRef]
57. Igbari, F.; Li, M.; Hu, Y.; Wang, Z.-K.; Liao, L.-S. A room-temperature $CuAlO_2$ hole interfacial layer for efficient and stable planar perovskite solar cells. *J. Mater. Chem. A* **2016**, *4*, 1326–1335. [CrossRef]
58. Mahmud, M.A.; Elumalai, N.K.; Upama, M.B.; Wang, D.; Gonçales, V.R.; Wright, M.; Gooding, J.J.; Haque, F.; Xu, C.; Uddin, A. Cesium compounds as interface modifiers for stable and efficient perovskite solar cells. *Sol. Energy Mater. Sol. Cells* **2018**, *174*, 172–186. [CrossRef]
59. Song, D.; Wei, D.; Cui, P.; Li, M.; Duan, Z.; Wang, T.; Ji, J.; Li, Y.; Mbengue, J.M.; Li, Y. Dual function interfacial layer for highly efficient and stable lead halide perovskite solar cells. *J. Mater. Chem. A* **2016**, *4*, 6091–6097. [CrossRef]
60. Kim, H.-S.; Lee, C.-R.; Im, J.-H.; Lee, K.-B.; Moehl, T.; Marchioro, A.; Moon, S.-J.; Humphry-Baker, R.; Yum, J.-H.; Moser, J.E. Lead iodide perovskite sensitized all-solid-state submicron thin film mesoscopic solar cell with efficiency exceeding 9%. *Sci. Rep.* **2012**, *2*, 591. [CrossRef] [PubMed]
61. Christians, J.A.; Miranda Herrera, P.A.; Kamat, P.V. Transformation of the excited state and photovoltaic efficiency of $CH_3NH_3PbI_3$ perovskite upon controlled exposure to humidified air. *J. Am. Chem. Soc.* **2015**, *137*, 1530–1538. [CrossRef] [PubMed]
62. Zhu, Z.; Bai, Y.; Liu, X.; Chueh, C.C.; Yang, S.; Jen, A.K.Y. Enhanced efficiency and stability of inverted perovskite solar cells using highly crystalline SnO_2 nanocrystals as the robust electron—Transporting layer. *Adv. Mater.* **2016**, *28*, 6478–6484. [CrossRef]
63. Aldibaja, F.K.; Badia, L.; Mas-Marzá, E.; Sánchez, R.S.; Barea, E.M.; Mora-Sero, I. Effect of different lead precursors on perovskite solar cell performance and stability. *J. Mater. Chem. A* **2015**, *3*, 9194–9200. [CrossRef]
64. Dkhissi, Y.; Weerasinghe, H.; Meyer, S.; Benesperi, I.; Bach, U.; Spiccia, L.; Caruso, R.A.; Cheng, Y.-B. Parameters responsible for the degradation of $CH_3NH_3PbI_3$-based solar cells on polymer substrates. *Nano Energy* **2016**, *22*, 211–222. [CrossRef]
65. Han, Y.; Meyer, S.; Dkhissi, Y.; Weber, K.; Pringle, J.M.; Bach, U.; Spiccia, L.; Cheng, Y.-B. Degradation observations of encapsulated planar $CH_3NH_3PbI_3$ perovskite solar cells at high temperatures and humidity. *J. Mater. Chem. A* **2015**, *3*, 8139–8147. [CrossRef]
66. Li, X.; Tschumi, M.; Han, H.; Babkair, S.S.; Alzubaydi, R.A.; Ansari, A.A.; Habib, S.S.; Nazeeruddin, M.K.; Zakeeruddin, S.M.; Grätzel, M. Outdoor performance and stability under elevated temperatures and long-term light soaking of triple-layer mesoporous perovskite photovoltaics. *Energy Technol.* **2015**, *3*, 551–555. [CrossRef]
67. Xie, F.X.; Zhang, D.; Su, H.; Ren, X.; Wong, K.S.; Graätzel, M.; Choy, W.C. Vacuum-assisted thermal annealing of $CH_3NH_3PbI_3$ for highly stable and efficient perovskite solar cells. *ACS Nano* **2015**, *9*, 639–646. [CrossRef]
68. Reese, M.O.; Gevorgyan, S.A.; Jørgensen, M.; Bundgaard, E.; Kurtz, S.R.; Ginley, D.S.; Olson, D.C.; Lloyd, M.T.; Morvillo, P.; Katz, E.A. Consensus stability testing protocols for organic photovoltaic materials and devices. *Sol. Energy Mater. Sol. Cells* **2011**, *95*, 1253–1267. [CrossRef]
69. Supasai, T.; Rujisamphan, N.; Ullrich, K.; Chemseddine, A.; Dittrich, T. Formation of a passivating CH3NH3PbI3/PbI2 interface during moderate heating of $CH_3NH_3PbI_3$ layers. *Appl. Phys. Lett.* **2013**, *103*, 183906. [CrossRef]
70. Shi, L.; Young, T.L.; Kim, J.; Sheng, Y.; Wang, L.; Chen, Y.; Feng, Z.; Keevers, M.J.; Hao, X.; Verlinden, P.J. Accelerated lifetime testing of organic–inorganic perovskite solar cells encapsulated by polyisobutylene. *ACS Appl. Mater. Interfaces* **2017**, *9*, 25073–25081. [CrossRef]
71. Elumalai, N.K.; Uddin, A. Hysteresis in organic-inorganic hybrid perovskite solar cells. *Sol. Energy Mater. Sol. Cells* **2016**, *157*, 476–509. [CrossRef]
72. Azpiroz, J.M.; Mosconi, E.; Bisquert, J.; De Angelis, F. Defect migration in methylammonium lead iodide and its role in perovskite solar cell operation. *Energy Environ. Sci.* **2015**, *8*, 2118–2127. [CrossRef]

73. Upama, M.B.; Elumalai, N.K.; Mahmud, M.A.; Wang, D.; Haque, F.; Gonçales, V.R.; Gooding, J.J.; Wright, M.; Xu, C.; Uddin, A. Role of fullerene electron transport layer on the morphology and optoelectronic properties of perovskite solar cells. *Org. Electron.* **2017**, *50*, 279–289. [CrossRef]
74. Kim, J.H.; Liang, P.W.; Williams, S.T.; Cho, N.; Chueh, C.C.; Glaz, M.S.; Ginger, D.S.; Jen, A.K.Y. High-performance and environmentally stable planar heterojunction perovskite solar cells based on a solution-processed copper-doped nickel oxide hole-transporting layer. *Adv. Mater.* **2015**, *27*, 695–701. [CrossRef] [PubMed]
75. Ono, L.K.; Raga, S.R.; Remeika, M.; Winchester, A.J.; Gabe, A.; Qi, Y. Pinhole-free hole transport layers significantly improve the stability of MAPbI$_3$-based perovskite solar cells under operating conditions. *J. Mater. Chem. A* **2015**, *3*, 15451–15456. [CrossRef]
76. Idígoras, J.; Aparicio, F.J.; Contreras-Bernal, L.; Ramos-Terrón, S.; Alcaire, M.; Sánchez-Valencia, J.R.; Borras, A.; Barranco, Á.; Anta, J.A. Enhancing moisture and water resistance in perovskite solar cells by encapsulation with ultrathin plasma polymers. *ACS Appl. Mater. Interfaces* **2018**, *10*, 11587–11594. [CrossRef] [PubMed]
77. Bella, F.; Griffini, G.; Correa-Baena, J.-P.; Saracco, G.; Grätzel, M.; Hagfeldt, A.; Turri, S.; Gerbaldi, C. Improving efficiency and stability of perovskite solar cells with photocurable fluoropolymers. *Science* **2016**, *354*, 203–206. [CrossRef] [PubMed]
78. Griffini, G.; Levi, M.; Turri, S. Novel crosslinked host matrices based on fluorinated polymers for long-term durability in thin-film luminescent solar concentrators. *Sol. Energy Mater. Sol. Cells* **2013**, *118*, 36–42. [CrossRef]
79. Griffini, G.; Levi, M.; Turri, S. Novel high-durability luminescent solar concentrators based on fluoropolymer coatings. *Prog. Organ. Coat.* **2014**, *77*, 528–536. [CrossRef]
80. Da Silva Sobrinho, A.; Latreche, M.; Czeremuszkin, G.; Klemberg-Sapieha, J.; Wertheimer, M. Transparent barrier coatings on polyethylene terephthalate by single-and dual-frequency plasma-enhanced chemical vapor deposition. *J. Vac. Sci. Technol. A Vac. Surf. Films* **1998**, *16*, 3190–3198. [CrossRef]
81. Da Silva Sobrinho, A.; Czeremuszkin, G.; Latreche, M.; Dennler, G.; Wertheimer, M. A study of defects in ultra-thin transparent coatings on polymers. *Surf. Coat. Technol.* **1999**, *116*, 1204–1210. [CrossRef]
82. Rossi, G.; Nulman, M. Effect of local flaws in polymeric permeation reducing barriers. *J. Appl. Phys.* **1993**, *74*, 5471–5475. [CrossRef]
83. Lee, Y.I.; Jeon, N.J.; Kim, B.J.; Shim, H.; Yang, T.Y.; Seok, S.I.; Seo, J.; Im, S.G. A low—Temperature thin—Film encapsulation for enhanced stability of a highly efficient perovskite solar cell. *Adv. Energy Mater.* **2018**, *8*, 1701928. [CrossRef]
84. Anderson, L.; Jacob, M. Effect of RF power on the optical and morphological properties of RF plasma polymerised linalyl acetate thin films. *Appl. Surf. Sci.* **2010**, *256*, 3293–3298. [CrossRef]
85. Xu, Q.F.; Wang, J.N.; Sanderson, K.D. Organic–inorganic composite nanocoatings with superhydrophobicity, good transparency, and thermal stability. *ACS Nano* **2010**, *4*, 2201–2209. [CrossRef] [PubMed]
86. Bazaka, K.; Jacob, M. Synthesis of radio frequency plasma polymerized non-synthetic Terpinen-4-ol thin films. *Mater. Lett.* **2009**, *63*, 1594–1597. [CrossRef]
87. Bazaka, K.; Jacob, M.V. Post-deposition ageing reactions of plasma derived polyterpenol thin films. *Polym. Degrad. Stab.* **2010**, *95*, 1123–1128. [CrossRef]
88. Jin, J.; Chen, S.; Zhang, J. Investigation of UV aging influences on the crystallization of ethylene-vinyl acetate copolymer via successive self-nucleation and annealing treatment. *J. Polym. Res.* **2010**, *17*, 827–836. [CrossRef]
89. Schlothauer, J.; Jungwirth, S.; Köhl, M.; Röder, B. Degradation of the encapsulant polymer in outdoor weathered photovoltaic modules: Spatially resolved inspection of EVA ageing by fluorescence and correlation to electroluminescence. *Sol. Energy Mater. Sol. Cells* **2012**, *102*, 75–85. [CrossRef]
90. Le Donne, A.; Dilda, M.; Crippa, M.; Acciarri, M.; Binetti, S. Rare earth organic complexes as down-shifters to improve Si-based solar cell efficiency. *Opt. Mater.* **2011**, *33*, 1012–1014. [CrossRef]
91. Hayes, R.A.; Lenges, G.M.; Pesek, S.C.; Roulin, J. Low Modulus Solar Cell Encapsulant Sheets with Enhanced Stability and Adhesion. United States Patent 8168885, 5 January 2012.
92. *Annual Report 2010*; Bayer AG: Leverkusen, Germany, 28 February 2011. Available online: https://www.bayer.com/en/gb-2010-en.pdfx (accessed on 29 November 2018).

93. Schut, J.H. Shining Opportunities in Solar Films. Plastics Technology. Available online: http://search.ebscohost.com/login.aspx (accessed on 29 November 2018).
94. Granstrom, J.; Swensen, J.; Moon, J.; Rowell, G.; Yuen, J.; Heeger, A. Encapsulation of organic light-emitting devices using a perfluorinated polymer. *Appl. Phys. Lett.* **2008**, *93*, 409. [CrossRef]
95. Cavalcante, L.M.; Schneider, L.F.J.; Silikas, N.; Watts, D.C. Surface integrity of solvent-challenged ormocer-matrix composite. *Dent. Mater.* **2011**, *27*, 173–179. [CrossRef]
96. Noller, K.; Vasko, K. Flexible polymer barrier films for the encapsulation of solar cells. In Proceedings of the 19th European Photovoltaic Solar Energy Conference and Exhibition 2004, Paris, France, 7–11 June 2004; pp. 2156–2159.
97. Budunoglu, H.; Yildirim, A.; Guler, M.O.; Bayindir, M. Highly transparent, flexible, and thermally stable superhydrophobic ORMOSIL aerogel thin films. *ACS Appl. Mater. Interfaces* **2011**, *3*, 539–545. [CrossRef]
98. Pagliaro, M.; Ciriminna, R.; Palmisano, G. Silica-based hybrid coatings. *J. Mater. Chem.* **2009**, *19*, 3116–3126. [CrossRef]
99. Fu, Y.; Tsai, F.-Y. Air-stable polymer organic thin-film transistors by solution-processed encapsulation. *Org. Electron.* **2011**, *12*, 179–184. [CrossRef]
100. Ong, K.S.; Raymond, G.C.R.; Ou, E.; Zheng, Z.; Ying, D.L.M. Interfacial and mechanical studies of a composite Ag–IZO–PEN barrier film for effective encapsulation of organic TFT. *Org. Electron.* **2010**, *11*, 463–466. [CrossRef]
101. Fluoropolymer. Available online: https://en.wikipedia.org/wiki/Fluoropolymer (accessed on 29 November 2018).

© 2019 by the authors. Licensee MDPI, Basel, Switzerland. This article is an open access article distributed under the terms and conditions of the Creative Commons Attribution (CC BY) license (http://creativecommons.org/licenses/by/4.0/).

Article

Long-Term Reliability Evaluation of Silica-Based Coating with Antireflection Effect for Photovoltaic Modules

Kensuke Nishioka [1,*], So Pyay Moe [1] and Yasuyuki Ota [1]

1. Faculty of Engineering, University of Miyazaki, Gakuen Kibanadai-nishi 1-1, Miyazaki 889-2192, Japan; spyay25@gmail.com
2. Organization for Promotion of Tenure Track, University of Miyazaki, Gakuen Kibanadai-nishi 1-1, Miyazaki 889-2192, Japan; y-ota@cc.miyazaki-u.ac.jp
* Correspondence: nishioka@cc.miyazaki-u.ac.jp; Tel.: +81-985-58-7774

Received: 8 December 2018; Accepted: 14 January 2019; Published: 15 January 2019

Abstract: Not all sunlight irradiated on the surface of a photovoltaic (PV) module can reach the cells in the PV module. This loss reduces the conversion efficiency of the PV module. The main factors of this loss are the reflection and soiling on the surface of the PV module. With this, it is effective to have both antireflection and antisoiling effects on the surface of PV modules. In this study, the antireflection and antisoiling effects along with the long-term reliability of the silica-based layer easily coated on PV modules were assessed. A silica-based layer with a controlled thickness and refractive index was coated on the surface of a $Cu(In,Ga)Se_2$ PV array. The array was exposed outdoors to assess its effects and reliability. As a result of the coating, the output of the PV array increased by 3.9%. The environment of the test site was relatively clean and the increase was considered to be a result of the antireflection effect. Moreover, it was observed that the effect of the coating was maintained without deterioration after 3.5 years. The coating was also applied to a silicon PV module and an effect similar to that of the CIGS PV module was observed in the silicon PV module.

Keywords: photovoltaic; coating; antireflection; antisoiling; long-term reliability

---

1. Introduction

With a rising problem in the field of energy, renewable energy sources that are non-reliant on fossil fuels continue to attract attention. Photovoltaics (PVs) was developed as an important and promising technology and has been a top eco-friendly energy solution among the various sustainable energy resources. Single and multi-crystalline silicon PVs are market leaders as first-generation PVs. Currently, thin-film PVs, such as compound semiconductor PVs, play a vital role [1,2].

Not all sunlight irradiated on the surface of the PV module reaches the cells. The loss, from incomplete utilization of sunlight, reduces the conversion efficiency of the PV module. Reflection and soiling on the surface of the PV module, References [3–6], are among the main factors for this loss. Both must be considered in the developing of high-efficiency PV systems. The cover glass of PV modules reflects approximately 4% of the incident light irradiated to it [7]. Energy losses of the incident light on PV modules occur through the reflection of the air/glass interface due to the difference in the refractive index of the medium [8]. As most PV systems are installed outdoors, it is inevitable for dust and other contaminants to be deposited on the system. Soiling on PV modules prevents light transmittance and decreases the performance of PV systems [9–13]. Therefore, it is relevant to incorporate antireflection and antisoiling effects on the PV module surface to improve the performance of PV systems.

To overcome common problems in PV systems, transparent coatings that have the ability to suppress reflection and soiling effects spark the interest in PV technologies. Coatings with high transmittance, and antireflection and self-cleaning properties gained significant attention among other types of coatings [14–16]. Silica-based coatings with a nano-sized porous structure have been recognized as a low-cost functional layer with antireflection and antisoiling effects [17,18]. By controlling its porosity, the refractive index can be adjusted to perform as an effective antireflection material on PV modules. For coatings, the refractive index and layer thickness were adjusted to increase transmittance [19]. Due to abundant hydroxyl groups in the silica coating, a very thin layer of water can be adsorbed onto the PV module surface. This can prevent localization of electrostatic charges by the electric conductivity of the thin water layer. The electrostatic charges can then be discharged to the environment. Thus, the adhesion of dust can be prevented [20]. Moreover, with the super-hydrophilicity of the coating, dust that adheres to the surface can easily be washed away with a minimal volume of washing water.

Another important key factor of PV systems is long-term reliability. This plays a paramount role in PV plants for a low-levelized cost of energy (LCOE). To retain high efficiency and low LCOE, soiling and reflection must be well-suppressed in PV systems for a long period of time. Practically speaking, it is important to investigate the long-term reliability of the coating on PV modules. This study assessed the antireflection and antisoiling effects along with the long-term reliability of the silica-based layer easily coated on PV modules.

2. Materials and Methods

In April 2010, a $Cu(In,Ga)Se_2$ (CIGS) PV system was installed on the roof of a seven-story building at the University of Miyazaki, Miyazaki, Japan. This system had never been cleaned until 25 December 2014, when cleaning and coating was carried out for the first time since installation. Three arrays (array A, array B, and array C) in the CIGS PV system were used in this study. Table 1 shows the details including the respective treatments applied to each CIGS array. All arrays comprised modules of the same model number and each, with a rated power of 10.2 kW, was composed of 120 modules. Array A was the reference array and did not undergo any treatment. Arrays B and C were cleaned with water using a rotating brush. After the cleaning, a silica-based layer with antireflection and antisoiling effects (EXCEL PURE, Central Automotive Products, Osaka, Japan) was coated on the module surface of array C. The dispersion medium of the coating liquid was composed of a mixture of alcohol and water, and silica having a size of 10 to 100 nm was dispersed.

Table 1. Details of each CIGS array.

Array	Number of Modules	Rated Power (kW)	Treatment
A	120	10.2	No (reference)
B	120	10.2	Cleaning
C	120	10.2	Cleaning and coating

The coating was carried out by immersing the coating liquid containing nano-sized silica in a sponge phase resin and scanning the resin on the PV modules, as shown in Figure 1a. Immediately after the coating, the solvent in the coating liquid evaporated and a porous silica layer was formed on the PV modules. The coated layer mainly comprised porous silica that contained many hydroxyl groups that adsorbed water on the surface. To confirm the thickness, the silica layer was coated on a glass substrate with the same method and a cross-sectional image was measured with a transmission electron microscope (JEM-2200FS, JEOL, Tokyo, Japan). The thickness of the coated layer was approximately 150 nm. Figure 1b shows the module surface after coating. The color of the coated area evidently turned darker. On 25 December 2014, cleaning and coating were carried out. The arrays were exposed outdoors to assess the effects and the reliability of the coating. Details of this treatment method were reported in Reference [19].

The array performance was evaluated using performance ratio (PR) as an indicator. PR was derived as follows:

$$PR = \frac{\frac{E_{out}(\text{kWh})}{GI\ (\text{kWh/m}^2)}}{\frac{10.2\ (\text{kW})\ [\text{rated power of a PV array}]}{1\ (\text{kW/m}^2)\ [\text{GI for rating}]}}, \quad (1)$$

where E_{out} and GI are the measured energy yield and amount of global solar radiation measured with a pyranometer, respectively.

Figure 1. (a) Coating method using sponge phase resin; (b) surface of the PV modules.

3. Results and Discussions

Although the three arrays investigated in this study comprised the same modules, there were some variations in the output between the arrays due to the differences of initial output and installation position. To cancel the initial output difference between the arrays, the daily PR was divided by the average PR for one year (from 25 December 2013, to 24 December 2014) before the treatment. Figure 2a shows the normalized daily PR (NPR) of Arrays B and C. NPRs for Arrays B (NPR_B) and C (NPR_C) showed seasonal variation. NPRs have higher values during the winter season (December, January, and February) and lower values during the summer season (July, August, and September). This is because during the summer and winter seasons, the module temperatures are high and low, respectively. NPR_B and NPR_C are almost the same value before the treatment (25 December 2014). On the contrary, NPR_C (cleaning and coating) after the treatment was clearly higher than NPR_B (cleaning only). In both arrays, a slight decrease in the PR, due to system degradation, was observed.

To clarify the effects of the treatment on the performance of the PV module, NPR_B (cleaning only) and NPR_C (cleaning and coating) were divided by the normalized daily PR of reference, array A (no treatment). Figure 2b shows the twice normalized daily PR of arrays B and C. In Figure 2b, the seasonal variation and age deterioration, which can be observed in Figure 2a, almost disappeared as a result of the second normalization. From this figure, the effect of the treatment can be clearly assessed. It was found that the twice normalized PR of array B before and after cleaning had little change. This result indicates that the cleaning did not increase the performance of the PV module at the installation site of this study. The environment of the test site was relatively clean because there were no industrial plants or arterial roads near the site and the arrays were installed at a high altitude (the rooftop of a seven-story building). In contrast, the twice normalized PR of array C increased after cleaning and coating. The effect of the coating was clearly observed and the performance of the PV module improved. As the test site was located in a clean environment and the antisoiling effect was not noticeable, this performance improvement was thought to be due to the antireflection effect.

To quantitatively evaluate the effect of the coating, NPR_C was divided by NPR_B. Figure 2c shows NPR_C/NPR_B. NPR_C/NPR_B immediately increased after the coating on 25 December 2014. The average value of NPR$_C$/NPR$_B$ from 25 December 2014, to 30 June 2018 was 1.039. The performance of the PV improved by 3.9% due to the antireflection effect of the coating.

Moreover, it was found that the effect of the coating was maintained without deterioration even after 3.5 years. In order to reduce the LCOE of PV systems, it is necessary to improve the long-term reliability of the elements constituting it. The coating can be applied via a simple process and with a film thickness as thin as 150 nm. Nevertheless, the coating maintained its effect, for 3.5 years, without deterioration.

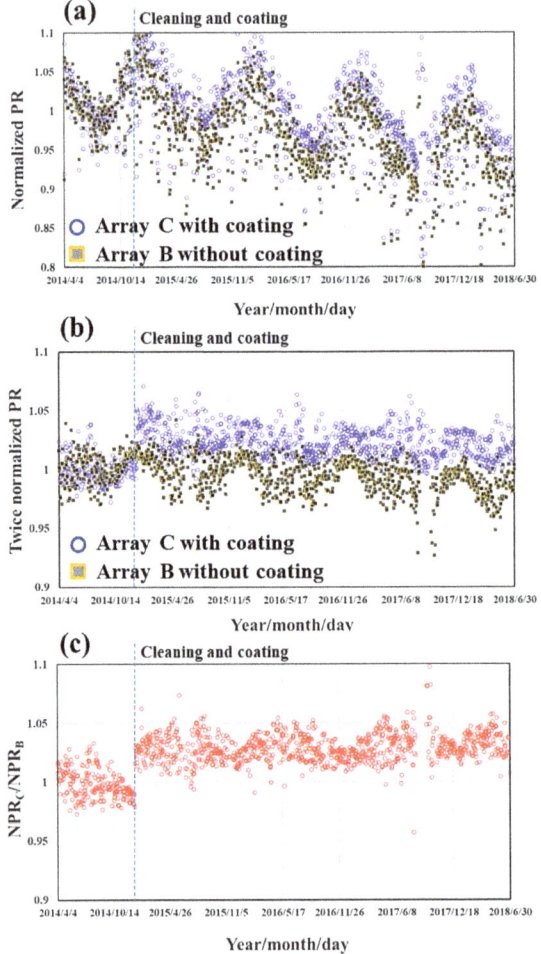

Figure 2. (a) Normalized daily PR of arrays B (NPR$_B$) and C (NPR$_C$), (b) twice normalized daily *PR* of arrays B and C (NPR$_B$ and NPR$_C$ were divided by the normalized daily *PR* of reference array A), and (c) NPR$_C$/NPR$_B$.

The main reason for reflection is the change in the refractive index from air (with a refractive index of 1) to glass (with a refractive index of 1.5 at 589.3 nm). In the case of PV modules, inserting a thin film with a refractive index between air and glass on the cover glass works as an antireflection layer.

Figure 3 shows the refractive index from 200 to 1200 nm of the coating assessed in this study. The refractive index was measured using a film thickness monitor (FE-3000, Otsuka Electronics,

Hirakata, Japan). The refractive index of the coating was 1.295 at 589.3 nm. This value is intermediate between air and glass which allowed the coating to act as an antireflection layer.

In this study, the silica-based coating was applied on the CIGS PV modules through a simple process and its effect on PV performance was assessed. The improvement of PV performance was maintained long term, 3.5 years without deterioration. This coating can be applied to other PV modules using glass materials as a cover glass.

Single-crystalline silicon PV modules of the same model number (GT85F, KIS, Saku, Japan) were prepared to be utilized for the assessment. The power of the modules was equal, rated at 90 W. The silica-based layer was coated on the surface of the module. Figure 4 shows the surface of the silicon PV modules after coating. The two modules were coated (modules X and X′) and the other two modules were not coated and act as references (modules Y and Y′). The color of the coated modules (modules X and X′) evidently changed to a darker shade. This was credited to the antireflection effect of the coating. The silicon PV modules were installed and assessed at the University of Miyazaki.

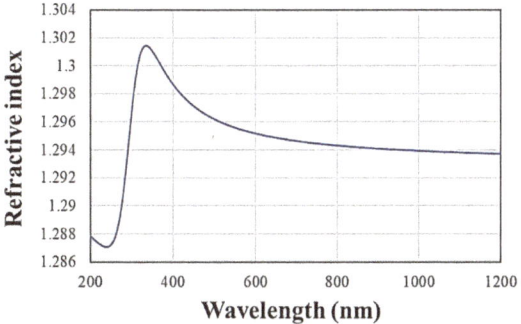

Figure 3. Refractive index as a function of the wavelength for the silica-based coating.

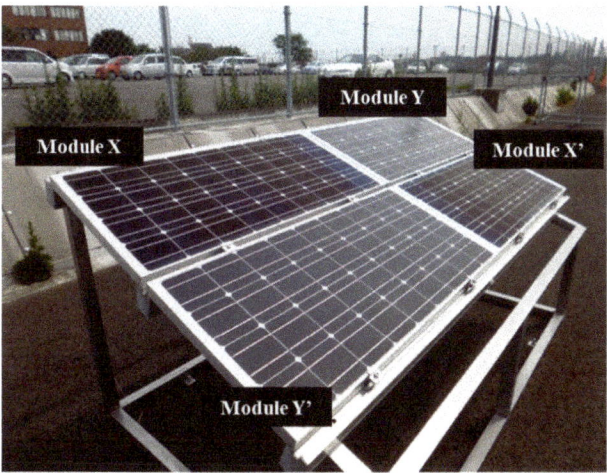

Figure 4. The surface of silicon PV modules. The two modules were coated (modules X and X′) and the other two modules were not coated to act as references (modules Y and Y′).

Figure 5 shows the current–voltage characteristics of module X before and after coating, measured using a solar simulator under standard test conditions. The short-circuit current increased, from 5.56 to 5.74 A (3.2% gain). The same coating effect as that of the CIGS PV module was also observed in the silicon PV module.

The single-crystalline silicon PV modules, with and without coating, were exposed outdoors from 1 December 2015 to assess the effect of the coating. To clarify the effect of the coating, the daily PR of module X, with coating (PR_X), was divided by that of module Y, without coating (PR_Y). Figure 6 shows the PR_X/PR_Y. The PR_X/PR_Y increased with an average value of 1.031 (a 3.1% gain) from 1 December 2015 to 1 April 2016. The environment in which the silicon PV modules were placed was found to be a clean environment with minimal dust and the output gain was considered to be due to the antireflection effect. The coating can be applied not only for CIGS PV modules but also for silicon PV modules.

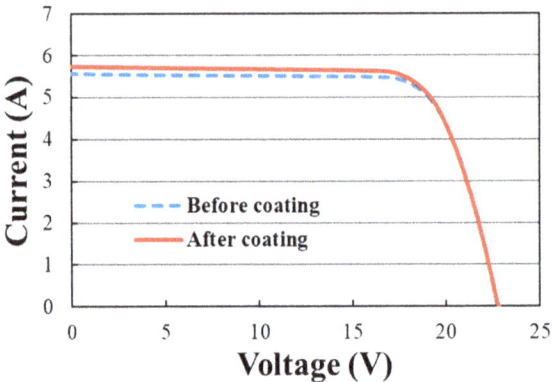

Figure 5. Current–voltage characteristics of module X before and after coating measured using a solar simulator under standard test conditions.

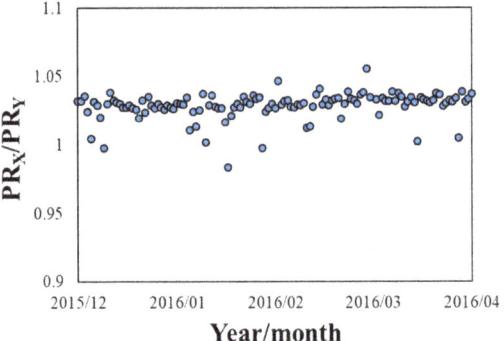

Figure 6. PR_X/PR_Y ratio versus time.

The cost to apply the coating (cleaning and coating) was 1.5 million JPY for a 1 MW PV system. The yearly output of a 1 MW PV system in Japan is approximately 1000 MWh. Assuming that the power selling price is 20 JPY/kWh and the output gain from the coating is 3.5%, an annual sales increase of 700,000 JPY can be expected. The coating fee will be paid in two years.

Many commercial PV power plants do periodic cleaning to keep energy yields up. The surface damage by the cleaning is becoming a big issue in the PV industry. The durability of the coating to the robot cleaning or other harsh cleaning has not been examined. However, it is expected that the frequency of cleaning and the harshness required for cleaning can be suppressed due to the antisoiling effect of the coating.

4. Conclusions

In this study, the antireflection and antisoiling effects, and long-term reliability of the silica-based layer easily coated on PV modules was assessed. A silica-based layer with antireflection and antisoiling effects was coated on the surface of CIGS PV arrays, and the arrays were exposed outdoors to assess the effects and the reliability of the coating. By controlling the porosity of the coating, the refractive index was adjusted to act as an effective antireflection material on PV modules. The output of the coated PV array showed a 3.9% increase in the amount of energy collected, confirming that the coating effectively increased the performance of the PV array due to the antireflection effects. Moreover, it was found that the effect of the coating was maintained without deterioration for more than 3.5 years from the date of coating. The environment of the test site was relatively clean, with a minimal amount of dust, as there were no industrial plants or arterial roads near the site location and the arrays were installed at a high altitude (on the rooftop of the seven-story building). The antisoiling effect of the coating was not considered as the PV array was not sufficiently dirty to exert a significant influence on the performance of the PV array. The coating was also applied to a silicon PV module and the same effect as that of the CIGS PV module was seen.

Author Contributions: Conceptualization, K.N.; Methodology, K.N. and Y.O.; Analysis, K.N., S.P.M. and Y.O.; Validation, K.N. and S.P.M.; Writing—Original draft preparation, K.N.; Writing—Review and editing, K.N.; Supervision, K.N.

Funding: This research received no external funding.

Acknowledgments: We would like to thank Satoshi Miyai of Central Automotive Products for valuable discussions.

Conflicts of Interest: The authors declare no conflict of interest.

References

1. Kushiya, K. CIS-based thin-film PV technology in solar frontier K.K. *Sol. Energy Mater. Sol. Cells* **2014**, *122*, 309–313. [CrossRef]
2. Kushiya, K. Key near-term R&D issues for continuous improvement in CIS-based thin-film PV modules. *Sol. Energy Mater. Sol. Cells* **2009**, *93*, 1037–1041. [CrossRef]
3. Elminir, H.K.; Ghitas, A.E.; Hamid, R.H.; El-Hussainy, F.; Beheary, M.M.; Abdel-Moneim, K.M. Effect of dust on the transparent cover of solar collectors. *Energy Convers. Manag.* **2006**, *47*, 3192–3203. [CrossRef]
4. García, M.; Marroyo, L.; Lorenzo, E.; Pérez, M. Soiling and other optical losses in solar-tracking PV plants in navarra. *Prog. Photovolt.* **2011**, *19*, 211–217. [CrossRef]
5. Piliougine, M.; Cañete, C.; Moreno, R.; Carretero, J.; Hirose, J.; Ogawa, S.; Sidrach-de-Cardona, M. Comparative analysis of energy produced by photovoltaic modules with anti-soiling coated surface in arid climates. *Appl. Energy* **2013**, *112*, 626–634. [CrossRef]
6. Paudyal, B.R.; Shakya, S.R. Dust accumulation effects on efficiency of solar PV modules for off grid purpose: A case study of Kathmandu. *Sol. Energy* **2016**, *135*, 103–110. [CrossRef]
7. Priyadarshini, B.G.; Sharma, A.K. Design of multi-layer anti-reflection coating for terrestrial solar panel glass. *Bull. Mater. Sci.* **2016**, *39*, 683–689. [CrossRef]
8. Arabatzis, I.; Todorova, N.; Fasaki, I.; Tsesmeli, C.; Peppas, A.; Li, W.X.; Zhao, Z. Photocatalytic, self-cleaning, antireflective coating for photovoltaic panels: Characterization and monitoring in real conditions. *Sol. Energy* **2018**, *159*, 251–259. [CrossRef]
9. El-Shobokshy, M.S.; Hussein, F.M. Effect of dust with different physical properties on the performance of photovoltaic cells. *Sol. Energy* **1993**, *51*, 505–511. [CrossRef]
10. El-Shobokshy, M.S.; Hussein, F.M. Degradation of photovoltaic cell performance due to dust deposition on to its surface. *Renew. Energy* **1993**, *3*, 585–590. [CrossRef]
11. Mastekbayeva, G.A.; Kumar, S. Effect of dust on the transmittance of low density polyethylene glazing in a tropical climate. *Sol. Energy* **2000**, *68*, 135–141. [CrossRef]
12. Kalogirou, S.A.; Agathokleous, R.; Panayiotou, G. On-site PV characterization and the effect of soiling on their performance. *Energy* **2013**, *51*, 439–446. [CrossRef]

13. Mani, M.; Pillai, M. Impact of dust on solar photovoltaic (PV) performance: Recent status, challenges and recommendations. *Renew. Sustain. Energy Rev.* **2010**, *14*, 3124–3131. [CrossRef]
14. Nabemoto, K.; Sakurada, Y.; Ota, Y.; Takami, K.; Nagai, H.; Tamura, K.; Araki, K.; Nishioka, K. Effect of anti-soiling layer coated on poly(methyl methacrylate) for concentrator photovoltaic modules. *Jpn. J. Appl. Phys.* **2012**, *51*, 10ND11. [CrossRef]
15. Sueto, T.; Ota, Y.; Nishioka, K. Suppression of dust adhesion on a concentrator photovoltaic module using an anti-soiling photocatalytic coating. *Sol. Energy* **2013**, *97*, 414–417. [CrossRef]
16. Hanaei, H.; Assadi, M.K.; Saidur, R. Highly efficient antireflective and self-cleaning coatings that incorporate carbon nanotubes (CNTs) into solar cells: A review. *Renew. Sustain. Energy Rev.* **2016**, *59*, 620–635. [CrossRef]
17. Lu, X.; Wang, Z.; Yang, X.; Xu, X.; Zhang, L.; Zhao, N.; Xu, J. Antifogging and antireflective silica film and its application on solar modules. *Surf. Coat. Technol.* **2011**, *206*, 1490–1494. [CrossRef]
18. Thompson, C.S.; Fleming, R.A.; Zou, M. Transparent self-cleaning and antifogging silica nanoparticle films. *Sol. Energy Mater. Sol. Cells* **2013**, *115*, 108–113. [CrossRef]
19. Ota, Y.; Ahmad, N.; Nishioka, K. A 3.2% output increase in an existing photovoltaic system using an antireflection and anti-soiling silica-based coat. *Sol. Energy* **2016**, *136*, 547–552. [CrossRef]
20. Horiuchi, Y.; Yamashita, H. Design of mesoporous silica thin films containing single-site photocatalysts and their applications to superhydrophilic materials. *Appl. Catal. A* **2011**, *400*, 1–8. [CrossRef]

© 2019 by the authors. Licensee MDPI, Basel, Switzerland. This article is an open access article distributed under the terms and conditions of the Creative Commons Attribution (CC BY) license (http://creativecommons.org/licenses/by/4.0/).

Review

Fabrication and SERS Performances of Metal/Si and Metal/ZnO Nanosensors: A Review

Grégory Barbillon

EPF-Ecole d' Ingenieurs, 3 bis rue Lakanal, 92330 Sceaux, France; gregory.barbillon@epf.fr

Received: 5 January 2019; Accepted: 28 January 2019; Published: 30 January 2019

Abstract: Surface-enhanced Raman scattering (SERS) sensors are very powerful analytical tools for the highly sensitive detection of chemical and biological molecules. Substantial efforts have been devoted to the design of a great number of hybrid SERS substrates such as silicon or zinc oxide nanosystems coated with gold/silver nanoparticles. By comparison with the SERS sensors based on Au and Ag nanoparticles/nanostructures, higher enhancement factors and excellent reproducibilities are achieved with hybrid SERS nanosensors. This enhancement can be due to the appearance of hotspots located at the interface between the metal (Au/Ag) and the semiconducting substrates. Thus, in this last decade, great advances in the domain of hybrid SERS nanosensors have occurred. In this short review, the recent advances of these hybrid metal-coated semiconducting nanostructures as SERS sensors of chemical and biological molecules are presented.

Keywords: SERS; sensors; plasmonics; silicon; zinc oxide; metals

1. Introduction

Since the 2000s, surface enhanced Raman scattering (SERS) spectroscopy based on a nanostructured substrate has become a powerful and promising technique for highly sensitive detection of chemical/biological molecules [1–4]. SERS performance depends on the substrates, which are mainly realized with metallic nanoparticles [5–9] or nanostructures [10–12] with different geometries obtained by various techniques such as X-ray interference lithography [13], interference lithography [14,15], deep UV lithography [16,17] and electron beam lithography [18–20]. Large SERS enhancement ability and reproducibility are achieved by these techniques. However, they require costly equipment and heavy manufacturing processes for mass production. Several low-cost strategies such as nanosphere lithography [21–23], nanoimprint lithography [24–26] and metallic nanoparticle assemblies by self-assembling processes [27–29] are employed as an alternative method for industrial production. Nonetheless, a limiting factor of these two low-cost techniques is the definition of nanostructures on large surfaces. More recently, alternative and promising SERS substrates have emerged and are composed of silicon nanowires associated with metallic nanoparticles [30–40] showing ultrahigh sensitivity and excellent reproducibility. For example, Galopin et al. investigated silicon nanowires coated with silver nanoparticles on large surfaces [38]. In addition, another alternative solution is to choose zinc oxide (ZnO) nanostructures. Indeed, this semiconducting material can be applied to SERS sensing [41–44] because ZnO has an excellent propensity for nanostructuration combined with a relatively high refractive index, which better confines the light and thus induces a potential enhancement of the SERS effect as for the silicon having also a high refractive index. For instance, Sinha et al. investigated ZnO nanorods coated with a gold layer [42]. Compared with SERS substrates only based on pure Au or Ag nanoparticles which have relatively low values of enhancement factors (EF), both hybrid nanostructures enable achieving higher EFs due to the hotspots located at the interface between the metal (Au/Ag) and the semiconducting substrates (Si/ZnO), which potentially improve the detection limit. Moreover, a great reproducibility of SERS signal is obtained with these hybrid nanosystems. Thus, a great number of hybrid nanostructures was carried out and employed as SERS substrates for obtaining high-performance sensors of different chemical/biological molecules with an excellent reliability and reproducibility.

During these two last decades, great progress has occurred in this research topic and a very small number of reviews concerning this progress can be found. Thus, an overview of these promising advances in the development of hybrid SERS nanosensors is presented. Firstly, we report on a review of the design and fabrication of hybrid SERS substrates based on silicon. Next, the representative results on the high sensitivity and the reproducibility of detection of chemical or biological molecules concerning tthese metal-coated Si SERS nanosensors are summarized. Lastly, a review on hybrid SERS substrates based on zinc oxide is reported concerning their fabrication and their SERS performances.

2. Hybrid SERS Nanosensors Based on Silicon Nanostructures

2.1. Fabrication Methods of Metal-Coated Si Nanosensors

Silicon nanostructures such as nanowires (SiNW) or nanopillars (SiNP) can mainly be obtained with the following fabrication methods, which are the solution phase synthesis (SPS) [45,46], the vapor–liquid–solid (VLS) growth [38,47–50], the oxide-assisted growth (OAG) [36,51,52], lithographic methods coupled to an etching process [53–55], and the metal-assisted chemical etching (MACE) [40,56,57] (see Figure 1a,b). More recently, a couple of techniques emerged such as maskless processes by reactive ion etching (RIE) [35] or RIE processes through the oxide native layer of silicon [58–60] (see Figure 1c). Moreover, a few groups employed a technique combining two existing techniques: the nanosphere lithography (NSL) and the MACE technique [61–64] (see Figure 1d for the principle scheme of this coupled technique of fabrication).

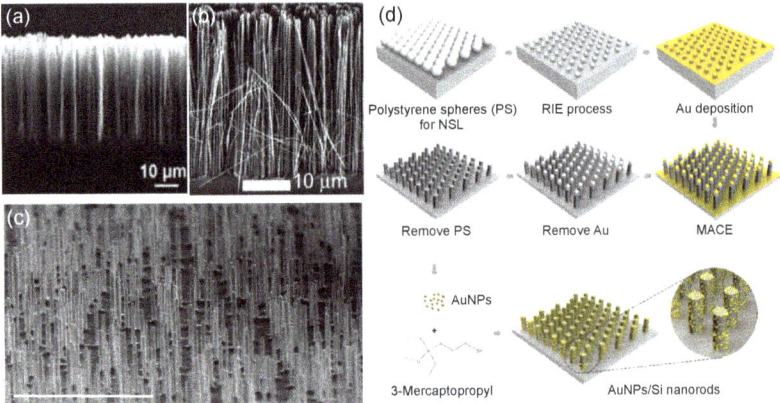

Figure 1. Three examples of metal-coated Si nanostructures (SEM images): (**a**) Si nanowires produced via a MACE technique (height $h = 24.3 \pm 1.9$ µm and nanowire diameter: 50 nm $< D <$ 300 nm) coated with Ag nanoparticles (average AgNP diameter: $d = 30$ nm), reproduced from [31] with permission from the Royal Society of Chemistry; (**b**) Si nanowires produced through an Ag-assisted chemical etching of n-Si(100) wafer (height $h \sim 32$ µm and diameter: 80 nm $< D <$ 200 nm), reprinted with permission from [37], Copyright 2010 American Chemical Society; (**c**) Au/Si nanopillars (NP) obtained with the fabrication technique based on an RIE process through the oxide native layer followed by a deposition of a gold layer of 30 nm (scale bar = 5 µm, radius $R \sim 125$ nm and height $h \sim 1000$ nm); (**d**) principle scheme of the low-cost fabrication process combining NSL and MACE for producing AuNP-conjugated Si nanorod arrays (reprinted with permission from [63], copyright 2017 American Chemical Society).

Thus, the SiNWs or SiNPs obtained with these various fabrication techniques are then modified with a metallic layer [35,58–60] or nanoparticles (e.g., Au and Ag nanoparticles) [31,36–39] in order to realize SERS substrates with significant performances in terms of the enhancement factor and the detection limit of biological and chemical molecules. The coating of Si nanostructures for SERS sensing is generally obtained with several Au nanostructures such as spherical nanoparticles, cylindrical nanorods, or triangular prisms [34]. Another technique employed for realizing metallic nanoparticles on silicon nanowires is the use of oxidation–reduction reactions, which consist of the reduction of metal ions by the electrons coming from the reaction on the SiNW surfaces etched by hydrofluoric acid [65]. These different metal-coated Si nanosensors have been fabricated in a major part at the wafer-scale [31,35,38,40,58–60,63].

2.2. SERS Performances of Metal-Coated Si Nanosystems for Chemical and Biological Sensing

In order to evaluate the SERS performances of metal-coated Si nanotructures for sensing of biological and chemical molecules, the enhancement factor (EF) is usually calculated. For that, two expressions are mainly employed. These two formulas giving EF and AEF (AEF = Analytical Enhancement Factor) are as follows:

$$EF = \frac{I_{SERS}}{I_{Raman}} \times \frac{N_{Raman}}{N_{SERS}}, \tag{1}$$

$$AEF = \frac{I_{SERS}}{I_{Raman}} \times \frac{C_{Raman}}{C_{SERS}}, \tag{2}$$

where I_{SERS} and I_{Raman} are SERS and Raman intensities for a given Raman peak, respectively. The N_{SERS} and N_{Raman} quantities are the numbers of excited molecules of analytes in SERS and Raman measurements, respectively. Finally, C_{SERS} and C_{Raman} correspond to concentrations of studied analytes used for SERS and Raman experiments, respectively. Firstly, the silicon nanowires coated with metallic nanoparticles demonstrated excellent SERS performances (see Table 1). For instance, Galopin et al. showed an EF factor in the range of 10^7–10^8 and a limit of detection (LOD) of 10 fM measured experimentally for the detection of Rhodamine 6G (R6G) molecules with Si nanowires coated with Ag nanoparticles [38] (see Figure 2a,b). With the same type of metal/Si systems (AgNPs/SiNWs), Zhang et al. obtained an EF factor in the range of 10^8–10^{10} for the detection of Sudan dyes [37]. Furthermore, an EF factor in the range of 6×10^9–8×10^9 is achieved and an LOD of 1 fM is experimentally measured for the DNA detection in the work of He et al. [36]. Besides, Wei et al. also demonstrated an LOD of 10 fM measured experimentally for the DNA sensing with Si nanowires coated with gold nanoparticles [33].

Secondly, the metal-coated Si nanosystems obtained with low-cost techniques also achieved great SERS performances for chemical and biological detection (see Table 1). Recently, Lin et al. fabricated Si nanorods with the coupling of the NSL and MACE techniques, and then coated these Si nanorods with gold nanoparticles (see Figures 1d and 2c,d). A value of 3×10^7 is achieved for the enhancement factor for the detection of Rhodamine 6G molecules [63] (see Figure 2d for SERS spectra). By using this technique of NSL associated with MACE, Cara et al. also showed a great EF of 1.6×10^6 for the detection of 7-mercapto-4-methylcoumarin (MMC) molecules with gold-coated Si nanowires [62]. With this same fabrication technique, Huang et al. achieved an EF of 1.1×10^6 for the detection of 4-aminothiophenol (4-AT) molecules with silver-coated Si nanowires [66]. By another way, Schmidt et al. demonstrated an EF value of 7×10^6 for the detection of trans-1,2-bis(4-pyridyl)ethylene (BPE) with leaning Ag-coated Si nanopillars. These nanosystems have been produced with a maskless process by using reactive ion etching. By leaning the Ag/Si nanopillars, hotspots are created and thus improving the Raman signal enhancement [35].

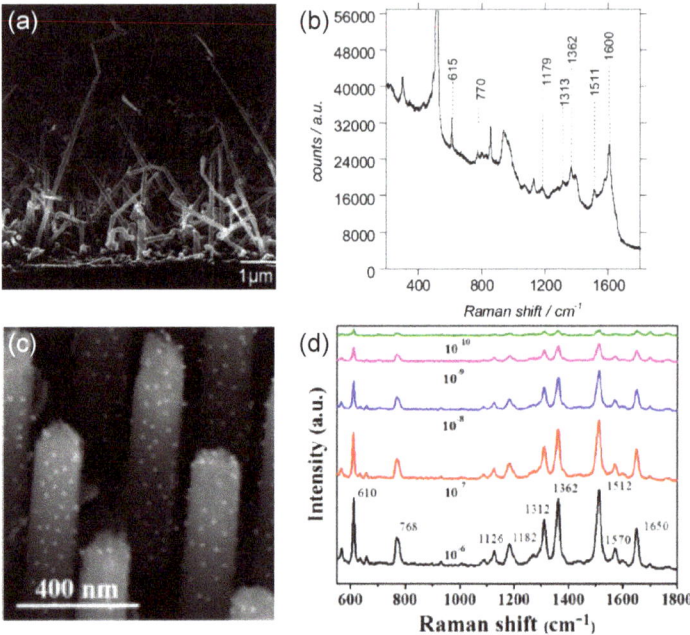

Figure 2. (**a**) Si nanowires produced by the vapor–liquid–solid growth (height: 5 µm < h < 6 µm and nanowire diameter: 50 nm < D < 150 nm) coated with Ag nanoparticles (diameter: 4 nm < d < 40 nm); (**b**) SERS spectrum of Rhodamine 6G (R6G) molecules ($C_{R6G} = 10^{-9}$ M) adsorbed on AgNPs/SiNWs. (**a**,**b**) are reprinted with permission from [38], copyright 2009 American Chemical Society; (**c**) Si nanorod array obtained with the combination of the NSL and MACE techniques followed by the grafting of gold nanoparticles having a diameter of 20 nm on Si nanorods (period $P \sim 500$ nm between two Si nanorods, diameter $D \sim 200$ nm and height $h \sim 1300$ nm); (**d**) SERS spectra of R6G molecules adsorbed on AuNPs/SiNRs for concentrations varying from 10^{-10} to 10^{-6} M; (**c**,**d**) are reprinted with permission from [63], copyright 2017 American Chemical Society.

Table 1. SERS performances of metal-coated Si nanosystems for chemical and biological sensing.

SERS Substrates	Detected Molecules	EF or AEF	LOD (M)	RSD	Refs.
AgNPs/SiNWs	R6G	10^7–10^8	10^{-14}	–	[38]
AgNPs/SiNWs	Sudan Dyes	10^8–10^{10}	–	<30%	[37]
AgNPs/SiNWs	MB	10^6–4×10^6	–	<10%	[31]
AgNPs/SiNWs	DNA	6×10^9–8×10^9	10^{-15}	–	[36]
Ag/SiNWs	Calcium Dipicolinate	–	4×10^{-6}	<20%	[40]
Ag/SiNWs	4-AT	1.1×10^6	–	<15%	[66]
AuNPs/SiNWs	DNA	–	10^{-14}	<10%	[33]
Au/SiNWs	MMC	1.6×10^6	–	<20%	[62]
AuNPs/SiNRs	R6G	3×10^7	10^{-10}	<8%	[63]
Leaning Ag/SiNPs	BPE	7×10^6	–	~10%	[35]
Au/SiNPs	Thiophenol	10^7–10^8	–	<7%	[58]
Al/SiNPs	Thiophenol	1.5×10^7–2.5×10^7	–	<7%	[60]

AgNPs = Ag nanoparticles; AuNPs = Au nanoparticles; Au or Ag or Al = metallic layer; SiNWs = Si nanowires; SiNRs = Si nanorods; SiNPs = Si nanopillars, and RSD = Relative Standard Deviation for SERS intensity.

In addition, Prof. Barbillon's group demonstrated excellent enhancement factors (EF or AEF) for the detection of thiophenol molecules with Au/Si and Al/Si nanopillars. For these two cases, Si nanopillars have been produced by a RIE process through the oxide native layer of silicon followed by the evaporation of a metallic layer. The values achieved for EF (for Au/Si) and AEF (for Al/Si) are in the range 10^7–10^8 [58] and 1.5×10^7–2.5×10^7 [60], respectively. Furthermore, this group also showed that the thickness of the metallic layer (for gold) has an effect on the SERS enhancement [59]. Lastly, other groups also demonstrated SERS enhancements with other shapes of Si nanosystems such as a tip-shaped silicon metasurface coated with gold nanoparticles [67,68], a hybrid Si nanospheroid network ornamented with gold nanospheres [69], silver-coated Si nanopores [70], or Si nanowire arrays coated silver nanoparticles [71,72]. To conclude Section 2, the interests of the metal-coated Si nanosystems are the use of fabrication techniques known as large-surface techniques, and also the plasmonic coupling between the metallic layer/nanoparticles and the semiconducting silicon substrate [38,58,73] in order to potentially improve the Raman enhancement. Moreover, a great reproducibility of the SERS signal is achieved with both metal-coated Si nanosystems produced with the techniques previously cited (average RSD = 15%, see Table 1). The Relative Standard Deviations for the SERS intensity of the studied Raman peaks are calculated on the basis of several SERS spectra recorded on different locations of each SERS substrate studied here. Nonetheless, some of these fabrication techniques are expensive and time consuming for a mass production. Finally, EF values obtained with the metal-coated Si nanosensors are higher or equal to more conventional SERS substrates such as core–shell nanosphere dimers (EF = 10^5–5×10^7 for Au–SiO$_2$ or Au–Pt nanosphere dimer) [74,75], a bare Au nanosphere dimer on a Si surface and on Pt surface (EF = 10^5–9×10^7) [74,76], or Au nanoparticles with a silver coating (EF = 10^4–10^5) [77].

3. Hybrid SERS Nanosensors Based on Zinc Oxide Nanostructures

3.1. Fabrication Methods of Metal-Coated ZnO Nanosensors

To produce metal-coated ZnO nanosystems, several techniques can be employed. At first, this begins by the fabrication of ZnO nanostructures by using various growth techniques such as the pulsed-laser deposition (PLD) [44,78–80], the hydrothermal growth [42,43,81] (see Figures 3a,b and 4a,c), the vapor–liquid–solid (VLS) or vapor–solid (VS) growth [41,82–84] (see Figure 5b). Then, after this step, metallic nanoparticles or a metallic layer are added with different techniques such as a sputtering process [41–43], an electron beam evaporation [44], a hydrothermal method [85], a controlled wet chemistry method associated with a spin-coater [86], a photochemical deposition method [81,87], and a dip-coating process [82]. In addition, a low-cost and wafer-scale technique based on NSL and solution processes has been developed by He et al. (see Figure 3d for the principle scheme of this fabrication method) for producing, for instance, urchin-like Ag nanoparticle/ZnO hollow nanosphere arrays (see Figure 3c) [88]. These different metal-coated ZnO nanosensors have been realized in a major part at the wafer-scale [41–43,85,86].

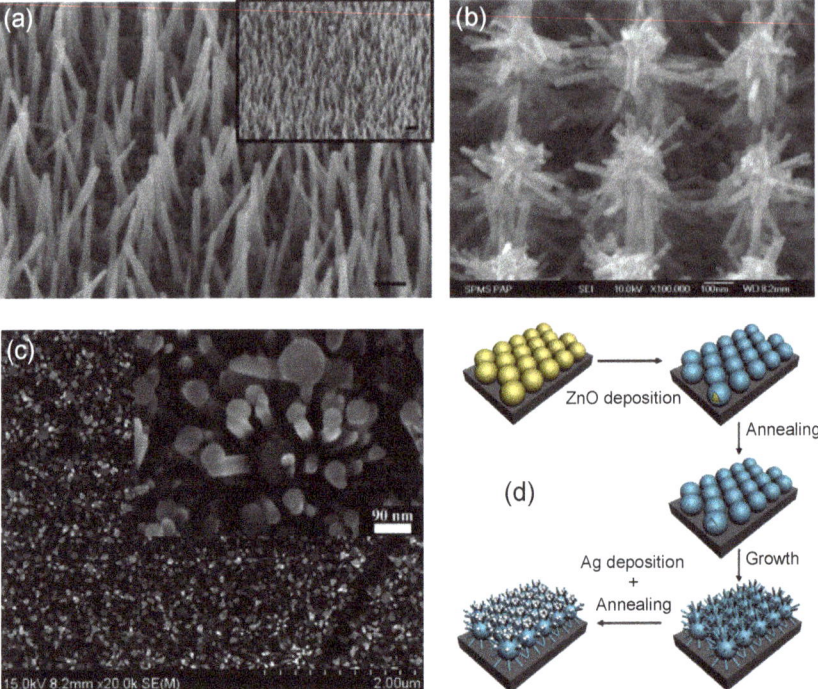

Figure 3. Three examples of metal-coated ZnO nanostructures: (**a**) SEM images of Au/ZnO nanorods and the insert is a picture at a weaker magnification (scale bars = 100 nm; reprinted with permission from [42], copyright 2011 American Chemical Society); (**b**) SEM image of AgNP-decorated Si/ZnO nanotrees (scale bar = 100 nm; reprinted with permission from [81], copyright 2010 American Chemical Society); (**c**) SEM images of an urchin-like AgNPs/ZnO hollow nanosphere (HNS) array and the insert corresponds to a picture recorded with a higher magnification; (**d**) principle scheme of the low-cost fabrication process of urchin-like AgNPs/ZnO HNS array; (**c**,**d**) are reproduced from [88] with permission from the Royal Society of Chemistry.

3.2. SERS Performances of Metal-Coated ZnO Nanosystems for Chemical Sensing

As previously, to evaluate the SERS performances of metal/ZnO nanotructures for sensing of chemical molecules, the enhancement factor (EF) is usually calculated by using one of the two equations Equations (1) and (2), and sometimes a formula of the SERS gain (G_{SERS}) is also used:

$$G_{SERS} = \frac{I_{SERS}}{I_{Raman}}, \qquad (3)$$

where I_{SERS} and I_{Raman} are SERS and Raman intensities for a given Raman peak, respectively. Firstly, the zinc oxide nanostructures coated with gold nanoparticles or a gold layer demonstrated excellent SERS performances (see Table 2). For instance, Chen et al. demonstrated an excellent EF of 1.2×10^7 obtained with Rhodamine 6G (R6G) molecules by using Au/ZnO nanoneedles (see Figure 4a,b). They also showed that the EF value depended on the density of gold nanoparticles and the morphology of Au/ZnO nanoneedles [85]. Another interesting example is those of Sinha et al. where zinc oxide nanorods have been coated with a gold layer, and a limit of detection (LOD) in terms of concentration was experimentally found and equals 1 pM for methylene blue (MB) molecules. They also demonstrated a good reproducibility of the SERS signal before each cleaning cycle assisted by UV [42]. Other groups observed different effects on the SERS signal such as the metal thickness [41], the density of gold/ZnO nanostructures [44] and the renewability of the gold/ZnO substrates [87,89]. Secondly, the zinc oxide nanostructures coated with Ag nanoparticles or a silver layer also showed great SERS performances (see Table 2). For example, Cui et al. showed a great EF evaluated at 2.5×10^{10} and an LOD of 1 pM measured experimentally for Malachite green (MG) molecules by using ZnO nanowire arrays coated with Ag nanoparticles (see Figure 4c,d). A low concentration of amoxicillin (1 nM) has been also experimentally detected with these Ag/ZnO nanostructures [43]. Furthermore, He et al. obtained great SERS performances for the detection of Rhodamine 6G molecules with urchin-like Ag nanoparticle/ZnO hollow nanosphere arrays. An EF of 10^8 is found and an LOD of 10^{-10} M is experimentally measured. They demonstrated that the SERS enhancement came from the great density of hotspots created by the multi-AgNP decoration, the charge transfer between Ag and ZnO, and the plasmonic coupling between AgNPs [88]. In another way, Song et al. observed a great EF of 1.2×10^8 and experimentally measured an LOD of 1 pM for p-aminothiophenol (PATP) molecules with ZnO nanofibers deposited on a silver foil surface. They demonstrated that SERS enhancement came from the exciton–plasmon interactions between ZnO nanofibers and the Ag foil surface, and also the fact that this type of Ag/ZnO nanosystem has a great photocatalytic activity towards pollutant (in this example, methylene blue (MB) molecules) degradation under UV illumination [90].

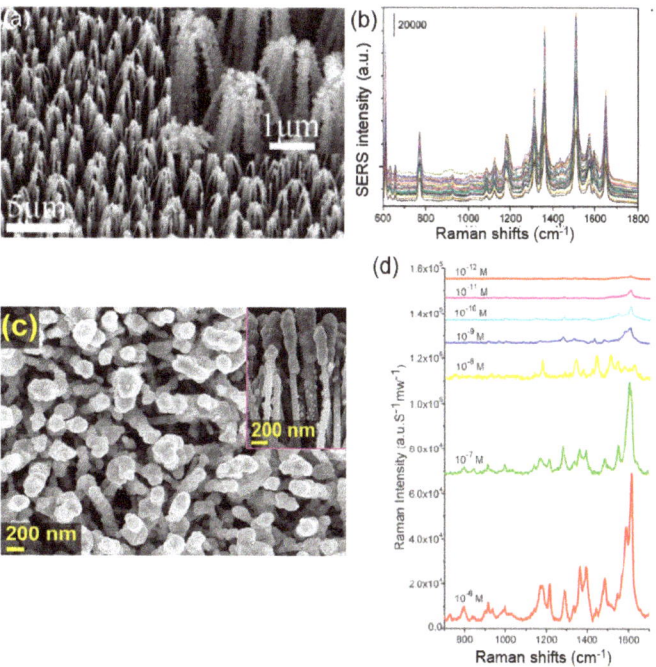

Figure 4. (a) SEM images of Au/ZnO nanoneedles and the insert is a zoom on a couple of nanoneedles; (b) SERS spectra of R6G molecules ($C_{R6G} = 10^{-7}$ M) adsorbed on Au/ZnO nanoneedles. (a,b) are reprinted with permission from [85], copyright 2010 American Chemical Society; (c) SEM images of Ag/ZnO nanowire array (NWA) and the insert corresponds to the cross-section of this array; (d) SERS spectra of MG molecules adsorbed on Ag/ZnO NWA for different MG concentrations; (c,d) are reproduced from [43] with permission from the Royal Society of Chemistry.

Table 2. SERS performances of metal-coated ZnO nanosystems for chemical sensing.

SERS Substrates	Detected Molecules	EF/AEF/G_{SERS}	LOD (M)	RSD	Refs.
Au/ZnO-NWs	4-MBT	2.2×10^6	–	<10%	[41]
Au/ZnO nanoneedles	R6G	1.2×10^7	–	<15%	[85]
Au/ZnO-NRs	MB	–	10^{-12}	–	[42]
Au/ZnO Inverse Nanostructures	Thiophenol	1.4×10^5	–	<20%	[89]
AuNPs/ZnO-NRs (Dendritic)	R6G	–	10^{-9}	–	[87]
Au/ZnO-NPs	Thiophenol	24	–	7%–45%	[44]
AuNPs/Ag/ZnO-Nanocones	Thiophenol	10^{10}–10^{11}	10^{-19}	–	[82]
AgNPs/ZnO-NWs	MG	2.5×10^{10}	10^{-12}	–	[43]
AgNPs/Si/ZnO nanotrees	R6G	10^6	–	<20%	[81]
AgNPs/wheatear-like H-ZnO	R6G	4.9×10^7	10^{-9}	<15%	[86]
AgFoil/ZnO nanofibers	PATP	1.2×10^8	10^{-12}	–	[90]
AgNPs/urchin-like ZnO-HNS	R6G	10^8	10^{-10}	<20%	[88]
AgNPs/ZnO-Nanoworms	R6G	3.1×10^7	10^{-10}	–	[91]
AgNPs/ZnO-NRs	4-ATP	2×10^6	–	–	[92]

AgNPs = Ag Nanoparticles; Au = metallic layer; ZnO-NWs = ZnO Nanowires; ZnO-NRs = ZnO Nanorods, H-ZnO = Hydrogenated ZnO, ZnO-NPs = ZnO Nanopillars, ZnO-HNS = ZnO Hollow Nanospheres, and RSD = Relative Standard Deviation for SERS intensity.

Several groups also demonstrated interesting properties for the Ag/ZnO nanosystems such as the superhydrophobicity [91], the hydrogenation of ZnO for improving the SERS enhancement [86], and their integration in microfluidic systems [92]. Finally, a couple of groups observed excellent EF with AgNPs/Si/ZnO and AuNPs/Ag/ZnO nanosystems. Indeed, Cheng et al. achieved an EF of 10^6 with Rhodamine 6G molecules by using AgNPs/Si/ZnO nanotrees [81], and Lee et al. an EF of 10^{10}–10^{11} and an LOD of 10^{-19} M which was experimentally measured with thiophenol molecules by using AuNPs/Ag/ZnO nanocones (see Figure 5). With this original design, Lee et al. obtained significant improvements of the SERS signal by using both following properties: the light trapping thanks to the nanocone arrays with a graded refractive index, the effect of plasmonic waveguide obtained with Ag-coated nanocones, and gap plasmons between AuNPs and Ag film (see Figure 5a) [82].

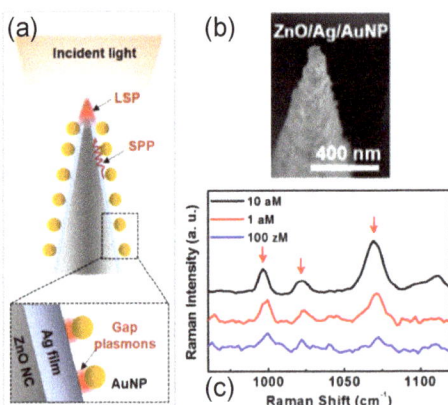

Figure 5. (a) scheme of the detection principle of SERS nanosensor (Electric field located at the tip and at the particle–film gap; SPP = Surface Plasmon Polariton and LSP = Localized Surface Plasmon); (b) SEM image of an AuNPs/Ag/ZnO nanocone; (c) SERS spectra of thiophenol molecules adsorbed on AuNPs/Ag/ZnO nanocones at the trace level (for three very low concentrations). All are reprinted with permission from [82], copyright 2015 American Chemical Society.

To conclude Section 3, the interests of the metal-coated ZnO nanosystems are mainly the use of low-cost fabrication techniques, the plasmonic coupling between the metallic layer/nanoparticles and the semiconducting substrate, and also plasmonic coupling of metallic nanoparticles in order to potentially improve the Raman enhancement. Moreover, a good reproducibility of the SERS signal is also achieved with both metal-coated ZnO nanosystems fabricated by the techniques previously cited (average RSD = 20%, see Table 2). As for metal-coated Si nanosystems, the RSD values concerning the SERS intensity of the studied Raman peaks are calculated on the basis of several SERS spectra recorded at different locations of each SERS substrate studied here. Finally, EF values obtained with the metal-coated ZnO nanosensors are higher or equal to more conventional SERS substrates such as core–shell nanosphere dimers (EF = 10^5–5×10^7 for Au–SiO$_2$ or Au–Pt nanosphere dimer) [74,75], a bare Au nanosphere dimer on a Si surface and on Pt surface (EF = 10^5–9×10^7) [74,76], or Au nanoparticles with a silver coating (EF = 10^4–10^5) [77].

4. Conclusions

In this short review, recent advances are presented concerning the development and the performances of SERS nanosensors based Si or ZnO nanostructures coated with a metallic layer or metallic nanoparticles for a very sensitive and reproducible detection of chemical and biological molecules. These metal-coated semiconducting SERS nanosensors highlighted excellent EF values and a great reproducibility with an average RSD of 15% and 20%, respectively, as described in the second and third sections. These two types of nanosensors can be produced with low-cost and wafer-scale techniques that have recently been developed. Thus, all these advances can enable an application of these SERS nanosensors to industrial and practical domains by using them as reliable, powerful and robust analytical platforms in the near future. However, a great number of works need to be conducted in order to better understand all the mechanisms involved in such metal-coated semiconducting nanostructures for enhancing the Raman signal. Furthermore, the hybrid metal-coated zinc oxide nanosystems can be used for applications to plasmonic photocatalysis thanks to the great photocatalytic efficiency of zinc oxide [93,94].

Funding: This research received no external funding.

Conflicts of Interest: The author declares no conflict of interest.

References

1. Yan, J.; Han, X.; He, J.; Kang, L.; Zhang, B.; Du, Y.; Zhao, H.; Dong, C.; Wang, H.-L.; Xu, P. Highly sensitive surface-enhanced Raman spectroscopy (SERS) platforms based on silver nanostructures fabricated on polyaniline membrane surfaces. *ACS Appl. Mater. Interfaces* **2012**, *4*, 2752–2756. [CrossRef] [PubMed]
2. Li, S.; Xu, P.; Ren, Z.; Zhang, B.; Du, Y.; Han, X.; Mack, N.H.; Wang, H.-L. Fabrication of thorny Au nanostructures on polyaniline surfaces for sensitive surface-enhanced Raman spectroscopy. *ACS Appl. Mater. Interfaces* **2013**, *5*, 49–54. [CrossRef] [PubMed]
3. Bodelon, G.; Montes-Garcia, V.; Lopez-Puente, V.; Hill, E.H.; Hamon, C.; Sanz-Ortiz, M.N.; Rodal-Cedeira, S.; Costas, C.; Celiksoy, S.; Perez-Juste, I.; et al. Detection and imaging of quorum sensing in *Pseudomonas aeruginosa* biofilm communities by surface-enhanced resonance Raman scattering. *Nat. Mater.* **2016**, *15*, 1203–1211. [CrossRef] [PubMed]
4. Sharma, B.; Cardial, M.F.; Kleinman, S.L.; Greeneltch, N.G.; Frontiera, R.R.; Blaber, M.G.; Schatz, G.C.; Van Duyne, R.P. High-performance SERS substrates: Advances and challenges. *MRS Bull.* **2013**, *38*, 615–624. [CrossRef]
5. Jimenez de Aberasturi, D.; Serano-Montes, A.B.; Langer, J.; Henriksen-Lacey, M.; Parak, W.J.; Liz-Marzan, L.M. Surface enhanced Raman scattering encoded gold nanostars for multiplexed cell discrimination. *Chem. Mater.* **2016**, *28*, 6779–6790. [CrossRef]
6. Rodriguez-Fernandez, D.; Langer, J.; Henriksen-Lacey, M.; Liz-Marzan, L.M. Hybrid Au-SiO$_2$ core-satellite colloids as switchable SERS tags. *Chem. Mater.* **2015**, *27*, 2540–2545. [CrossRef]

7. La Porta, A.; Sanchez-Iglesias, A.; Altantzis, T.; Bals, S.; Grzelczak, M.; Liz-Marzan, L.M. Multifunctional self-assembled composite colloids and their application to SERS detection. *Nanoscale* **2015**, *7*, 10377–10381. [CrossRef]
8. Wang, Q.; Lu, G.; Hou, L.; Zhang, T.; Luo, C.; Yang, H.; Barbillon, G.; Lei, F.H.; Marquette, C.A.; Perriat, P.; et al. Fluorescence correlation spectroscopy near individual gold nanoparticle. *Chem. Phys. Lett.* **2011**, *503*, 256–261. [CrossRef]
9. Dalstein, L.; Ben Haddada, M.; Barbillon, G.; Humbert, C.; Tadjeddine, A.; Boujday, S.; Busson, B. Revealing the interplay between adsorbed molecular layers and gold nanoparticles by linear and nonlinear optical properties. *J. Phys. Chem. C* **2015**, *119*, 17146–17155. [CrossRef]
10. Yu, Q.; Guan, P.; Qin, D.; Golden, G.; Wallace, P.M. Inverted size-dependence of surface-enhanced Raman scattering on gold nanohole and nanodisk arrays. *Nano Lett.* **2008**, *8*, 1923–1928. [CrossRef]
11. Faure, A.-C.; Barbillon, G.; Ou, M.; Ledoux, G.; Tillement, O.; Roux, S.; Fabregue, D.; Descamps, A.; Bijeon, J.-L.; Marquette, C.A.; et al. Core/shell nanoparticles for multiple biological detection with enhanced sensitivity and kinetics. *Nanotechnology* **2008**, *19*, 485103. [CrossRef] [PubMed]
12. Barbillon, G.; Faure, A.-C.; El Kork, N.; Moretti, P.; Roux, S.; Tillement, O.; Ou, M.G.; Descamps, A.; Perriat, P.; Vial, A.; et al. How nanoparticles encapsulating fluorophores allow a double detection of biomolecules by localized surface plasmon resonance and luminescence. *Nanotechnology* **2008**, *19*, 035705. [CrossRef] [PubMed]
13. Zhang, P.; Yang, S.; Wang, L.; Zhao, J.; Zhu, Z.; Liu, B.; Zhong, J.; Sun, X. Large-scale uniform Au nanodisk arrays fabricated via X-ray interference lithography for reproducible and sensitive SERS substrate. *Nanotechnology* **2014**, *25*, 245301. [CrossRef] [PubMed]
14. Barbillon, G.; Bijeon, J.-L.; Lérondel, G.; Plain, J.; Royer, P. Detection of chemical molecules with integrated plasmonic glass nanotips. *Surf. Sci.* **2008**, *602*, L119–L122. [CrossRef]
15. Ahn, H.-J.; Thiyagarajan, P.; Jia, L.; Kim, S.-I.; Yoon, J.-C.; Thomas, E.L.; Jang, J.-H. An optimal substrate design for SERS: Dual-scale diamond-shaped gold nano-structures fabricated via interference lithography. *Nanoscale* **2013**, *5*, 1836–1842. [CrossRef]
16. Vo-Dinh, T.; Dhawan, A.; Norton, S.J.; Khoury, C.G.; Wang, H.N.; Misra, V.; Gerhold, M.D. Plasmonic nanoparticles and nanowires: Design, fabrication and application in sensing. *J. Phys. Chem. C* **2010**, *114*, 7480–7488. [CrossRef] [PubMed]
17. Dhawan, A.; Duval, A.; Nakkach, M.; Barbillon, G.; Moreau, J.; Canva, M.; Vo-Dinh, T. Deep UV nano-microstructuring of substrates for surface plasmon resonance imaging. *Nanotechnology* **2011**, *22*, 165301. [CrossRef]
18. Gillibert, R.; Sarkar, M.; Bryche, J.-F.; Yasukuni, R.; Moreau, J.; Besbes, M.; Barbillon, G.; Bartenlian, B.; Canva, M.; Lamy de la Chapelle, M. Directional surface enhanced Raman scattering on gold nano-gratings. *Nanotechnology* **2016**, *27*, 115202. [CrossRef]
19. Bryche, J.-F.; Gillibert, R.; Barbillon, G.; Sarkar, M.; Coutrot, A.-L.; Hamouda, F.; Aassime, A.; Moreau, J.; Lamy de la Chapelle, M.; Bartenlian, B.; et al. Density effect of gold nanodisks on the SERS intensity for a highly sensitive detection of chemical molecules. *J. Mater. Sci.* **2015**, *50*, 6601–6607. [CrossRef]
20. Bryche, J.-F.; Gillibert, R.; Barbillon, G.; Gogol, P.; Moreau, J.; Lamy de la Chapelle, M.; Bartenlian, B.; Canva, M. Plasmonic enhancement by a continuous gold underlayer: Application to SERS sensing. *Plasmonics* **2016**, *11*, 601–608. [CrossRef]
21. Bryche, J.-F.; Tsigara, A.; Bélier, B.; Lamy de la Chapelle, M.; Canva, M.; Bartenlian, B.; Barbillon, G. Surface enhanced Raman scattering improvement of gold triangular nanoprisms by a gold reflective underlayer for chemical sensing. *Sens. Actuator B* **2016**, *228*, 31–35. [CrossRef]
22. Masson, J.-F.; Gibson, K.F.; Provencher-Girard, A. Surface-enhanced Raman spectroscopy amplification with film over etched nanospheres. *J. Phys. Chem. C* **2010**, *114*, 22406–22412. [CrossRef]
23. Camden, J.P.; Dieringer, J.A.; Zhao, J.; Van Duyne, R.P. Controlled plasmonic nanostructures for surface-enhanced spectroscopy and sensing. *Acc. Chem. Res.* **2008**, *41*, 1653–1661. [CrossRef] [PubMed]
24. Hamouda, F.; Sahaf, H.; Held, S.; Barbillon, G.; Gogol, P.; Moyen, E.; Aassime, A.; Moreau, J.; Canva, M.; Lourtioz, J.-M.; et al. Large area nanopatterning by combined anodic aluminum oxide and soft UV-NIL technologies for applications in biology. *Microelectron. Eng.* **2011**, *88*, 2444–2446. [CrossRef]

25. Cottat, M.; Lidgi-Guigui, N.; Tijunelyte, I.; Barbillon, G.; Hamouda, F.; Gogol, P.; Aassime, A.; Lourtioz, J.-M.; Bartenlian, B.; Lamy de la Chapelle, M. Soft UV nanoimprint lithography-designed highly sensitive substrates for SERS detection. *Nanoscale Res. Lett.* **2014**, *9*, 623. [CrossRef] [PubMed]
26. Jahn, M.; Patze, S.; Hidi, I.J.; Knipper, R.; Radu, A.I.; Mühling, A.; Yüksel, S.; Peksa, V.; Weber, K.; Mayerhöfer, T.; et al. Plasmonic nanostructures for surface enhanced spectroscopic methods. *Analyst* **2016**, *141*, 756–793. [CrossRef] [PubMed]
27. Que, R.; Shao, M.; Zhuo, S.; Wen, C.; Wang, S.; Lee, S.-T. Highly reproducible surface-enhanced Raman scattering on a capillarity-assisted gold nanoparticle assembly. *Adv. Funct. Mater.* **2011**, *21*, 3337–3343. [CrossRef]
28. Tanoue, Y.; Sugawa, K.; Yamamuro, T.; Akiyama, T. Densely arranged two-dimensional silver nanoparticle assemblies with optical uniformity over vast areas as excellent surface-enhanced Raman scattering substrates. *Phys. Chem. Chem. Phys.* **2013**, *15*, 15802–15805. [CrossRef]
29. Sugawa, K.; Akiyama, T.; Tanoue, Y.; Harumoto, T.; Yanagida, S.; Yasumori, A.; Tomita, S.; Otsuki, J. Particle size dependence of the surface-enhanced Raman scattering properties of densely arranged two-dimensional assemblies of Au(core)–Ag(shell) nanospheres. *Phys. Chem. Chem. Phys.* **2015**, *17*, 21182–21189.
30. Convertino, A.; Mussi, V.; Maiolo, L. Disordered array of Au covered silicon nanowires for SERS biosensing combined with electrochemical detection. *Sci. Rep.* **2016**, *6*, 25099. [CrossRef]
31. Akin, M.S.; Yilmaz, M.; Babur, E.; Ozdemur, B.; Erdogan, H.; Tamer, U.; Demirel, G. Large area uniform deposition of silver nanoparticles through bio-inspired polydopamine coating on silicon nanowire arrays for pratical SERS applications. *J. Mater. Chem. B* **2014**, *2*, 4894–4900. [CrossRef]
32. Wang, H.; Jiang, X.; Lee, S.T.; He, Y. Silicon nanohybrid-based surface-enhanced Raman scattering sensors. *Small* **2014**, *10*, 4455–4468. [CrossRef] [PubMed]
33. Wei, X.P.; Su, S.; Guo, Y.Y.; Jiang, X.X.; Zhong, Y.L.; Su, Y.Y.; Fan, C.H.; Lee, S.T.; He, Y. A molecular beacon-based signal-off surface-enhanced Raman scattering strategy for highly sensitive, reproducible, and multiplexed DNA detection. *Small* **2013**, *9*, 2493–2499. [CrossRef] [PubMed]
34. Chen, R.; Li, D.; Hu, H.; Zhao, Y.; Wang, Y.; Wong, N.; Wang, S.; Zhang, Y.; Hu, J.; Shen, Z.; et al. Tailoring optical properties of silicon nanowires by Au nanostructure decorations: Enhanced Raman scattering and photodetection. *J. Phys. Chem. C* **2012**, *116*, 4416–4422. [CrossRef]
35. Schmidt, M.S.; Hübner, J.; Boisen, A. Large area fabrication of leaning silicon nanopillars for surface enhanced Raman spectroscopy. *Adv. Mater.* **2012**, *24*, OP11–OP18. [CrossRef] [PubMed]
36. He, Y.; Su, S.; Xu, T.T.; Zhong, Y.L.; Zapien, J.A.; Li, J.; Fan, C.H.; Lee, S.T. Silicon nanowires-based highly-efficient SERS-active platform for ultrasensitive DNA detection. *Nano Today* **2011**, *6*, 122–130. [CrossRef]
37. Zhang, M.L.; Fan, X.; Zhou, H.W.; Shao, M.W.; Antonio Zapien, J.; Wong, N.B.; Lee, S.T. A high-efficiency surface-enhanced Raman scattering substrate based on silicon nanowires array decorated with silver nanoparticles. *J. Phys. Chem. C* **2010**, *114*, 1969–1975. [CrossRef]
38. Galopin, E.; Barbillat, J.; Coffinier, Y.; Szunerits, S.; Patriarche, G.; Boukherroub, R. Silicon nanowires coated with silver nanostructures as ultrasensitive interfaces for surface-enhanced Raman spectroscopy. *ACS Appl. Mater. Interfaces* **2009**, *1*, 1396–1403. [CrossRef]
39. Chen, J.; Martensson, T.; Dick, K.A.; Deppert, K.; Xu, H.Q.; Samuelson, L.; Xu, H. Surface-enhanced Raman scattering of rhodamine 6G on nanowire arrays decorated with gold nanoparticles. *Nanotechnology* **2008**, *19*, 275712. [CrossRef]
40. Zhang, B.H.; Wang, H.S.; Lu, L.H.; Ai, K.L.; Zhang, G.; Cheng, X.L. Large-area silver-coated silicon nanowire arrays for molecular sensing using surface-enhanced Raman spectroscopy. *Adv. Funct. Mater.* **2008**, *18*, 2348–2355. [CrossRef]
41. Khan, M.A.; Hogan, T.P.; Shanker, B. Gold-coated zinc oxide nanowire-based substrate for surface-enhanced Raman spectroscopy. *J. Raman Spectrosc.* **2009**, *40*, 1539–1545. [CrossRef]
42. Sinha, G.; Depero, L.E.; Alessandri, I. Recyclabe SERS substrates based on Au-coated ZnO nanorods. *ACS Appl. Mater. Interfaces* **2011**, *3*, 2557–2563. [CrossRef] [PubMed]
43. Cui, S.; Dai, Z.; Tian, Q.; Liu, J.; Xiao, X.; Jiang, C.; Wu, W.; Roy, V.A.L. Wetting properties and SERS applications of ZnO/Ag nanowire arrays patterned by a screen printing method. *J. Mater. Chem. C* **2016**, *4*, 6371–6379. [CrossRef]
44. Barbillon, G.; Sandana, V.E.; Humbert, C.; Bélier, B.; Rogers, D.J.; Teherani, F.H.; Bove, P.; McClintock, R.; Razeghi, M. Study of Au coated ZnO nanoarrays for surface enhanced Raman scattering chemical sensing. *J. Mater. Chem. C* **2017**, *5*, 3528–3535. [CrossRef]

45. Holmes, J.D.; Johnston, K.P.; Doty, R.C.; Korgel, B.A. Control of thickness and orientation of solution-grown silicon nanowires. *Science* **2000**, *287*, 1471–1473. [CrossRef]
46. Tuan, H.Y.; Lee, D.C.; Hanrath, T.; Korgel, B.A. Catalytic solid-phase seeding of silicon nanowires by nickel nanocrystals in organic solvents. *Science* **2005**, *5*, 681–684. [CrossRef] [PubMed]
47. Hochbaum, A.I.; Fan, R.; He, R.; Yang, P. Controlled growth of Si nanowire arrays for device integration. *Nano Lett.* **2005**, *5*, 457–460. [CrossRef]
48. Kayes, B.M.; Filler, M.A.; Putnam, M.C.; Kelzenberg, M.D.; Lewis, N.S.; Atwater, H.A. Growth of vertically aligned Si wire arrays over large areas (>1 cm^2) with Au and Cu catalysts. *Appl. Phys. Lett.* **2007**, *91*, 103110. [CrossRef]
49. Putnam, M.C.; Filler, M.A.; Kayes, B.M.; Kelzenberg, M.D.; Guan, Y.; Lewis, N.S.; Eiler, J.M.; Atwater, H.A. Secondary ion mass spectrometry of vapor–liquid–solid grown, Au-catalyzed, Si wires. *Nano Lett.* **2008**, *8*, 3109–3113. [CrossRef]
50. Schmidt, V.; Wittemann, J.; Gosele, U. Growth, thermodynamics, and electrical properties of silicon nanowires. *Chem. Rev.* **2010**, *110*, 361–388. [CrossRef]
51. Lee, S.T.; Wang, N.; Zhang, Y.F.; Tang, Y.H. Oxide-assisted semiconductor nanowire growth. *MRS Bull.* **1999**, *24*, 36–42. [CrossRef]
52. Ma, D.D.D.; Lee, C.S.; Au, F.C.K.; Tong, S.Y.; Lee, S.T. Small-diameter silicon nanowire surfaces. *Science* **2003**, *299*, 1874–1877. [CrossRef] [PubMed]
53. Caldwell, J.D.; Glembocki, O.; Bezares, F.J.; Bassim, N.D.; Rendell, R.W.; Feygelson, M.; Ukaegbu, M.; Kasica, R.; Shirey, L.; Hosten, C. Plasmonic nanopillar arrays for large-area, high-enhancement surface-enhanced Raman scattering sensors. *ACS Nano* **2011**, *5*, 4046–4055. [CrossRef] [PubMed]
54. Juhasz, R.; Elfstrom, N.; Linnros, J. Controlled fabrication of silicon nanowires by electron beam lithography and electrochemical size reduction. *Nano Lett.* **2005**, *5*, 275–280. [CrossRef] [PubMed]
55. Khorasaninejad, M.; Walia, J.; Saini, S.S. Enhanced first-order Raman scattering from arrays of vertical silicon nanowires. *Nanotechnology* **2012**, *23*, 275706. [CrossRef] [PubMed]
56. Peng, K.Q.; Hu, J.J.; Yan, Y.J.; Wu, Y.; Fang, H.; Xu, Y.; Lee, S.T.; Zhu, J. Fabrication of single-crystalline silicon nanowires by scratching a silicon surface with catalytic metal particles. *Adv. Funct. Mater.* **2006**, *16*, 387–394. [CrossRef]
57. Huang, Z.; Geyer, N.; Werner, P.; de Boor, J.; Gösele, U. Metal-assisted chemical etching of silicon: A review. *Adv. Mater.* **2011**, *23*, 285–308. [CrossRef] [PubMed]
58. Bryche, J.-F.; Bélier, B.; Bartenlian, B.; Barbillon, G. Low-cost SERS substrates composed of hybrid nanoskittles for a highly sensitive sensing of chemical molecules. *Sens. Actuator B* **2017**, *239*, 795–799. [CrossRef]
59. Magno, G.; Bélier, B.; Barbillon, G. Gold thickness impact on the enhancement of SERS detection in low-cost Au/Si nanosensors. *J. Mater. Sci.* **2017**, *52*, 13650–13656. [CrossRef]
60. Magno, G.; Bélier, B.; Barbillon, G. Al/Si nanopillars as very sensitive SERS substrates. *Materials* **2018**, *11*, 1534. [CrossRef]
61. Kara, S.A.; Keffous, A.; Giovannozzi, A.M.; Rossi, A.M.; Cara, E.; D'Ortenzi, L.; Sparnacci, K.; Boarino, L.; Gabouze, N.; Soukane, S. Fabrication of flexible silicon nanowires by sefl-assembled metal assisted chemical etching for surface enhanced Raman spectroscopy. *RSC Adv.* **2016**, *6*, 93649–93659. [CrossRef]
62. Cara, E.; Mandrile, L.; Lupi, F.F.; Giovannozzi, A.M.; Dialameh, M.; Portesi, C.; Sparnacci, K.; De Leo, N.; Rossi, A.M.; Boarino, L. Influence of the long-range ordering of gold-coated Si nanowires on SERS. *Sci. Rep.* **2018**, *8*, 11305. [CrossRef] [PubMed]
63. Lin, D.; Wu, Z.; Li, S.; Zhao, W.; Ma, C.; Wang, J.; Jiang, Z.; Zhong, Z.; Zheng, Y.; Yang, X. Large-area Au-nanoparticle-functionalized Si nanorod arrays for spatially uniform surface-enhanced Raman spectroscopy. *ACS Nano* **2017**, *11*, 1478–1487. [CrossRef] [PubMed]
64. Liang, Z.; Fan, D. Visible light-gated reconfigurable rotary actuation of electric nanomotors. *Sci. Adv.* **2018**, *4*, eaau0981. [CrossRef] [PubMed]
65. Peng, K.Q.; Lu, A.J.; Wong, N.B.; Zhang, R.Q.; Lee, S.T. Motility of metal nanoparticles in silicon and induced anisotropic silicon etching. *Adv. Funct. Mater.* **2008**, *18*, 3026–3035. [CrossRef]
66. Huang, J.-A.; Zhao, Y.-Q.; Zhang, X.-J.; He, L.-F.; Wong, T.-L.; Chui, Y.-S.; Zhang W.-J.; Lee, S.-T. Ordered Ag/Si nanowires array: Wide-range surface-enhanced Raman spectroscopy for reproducible biomolecule detection. *Nano Lett.* **2013**, *13*, 5039–5045. [CrossRef] [PubMed]

67. Lagarkov, A.; Boginkaya, I.; Bykov, I.; Budashov, I.; Ivanov, A.; Kurochkin, I.; Ryzhikov, I.; Rodionov, I.; Sedova, M.; Zverev, A.; et al. Light localization and SERS in tip-shaped silicon metasurface. *Opt. Express* **2017**, *25*, 17021–17038. [CrossRef]
68. Sarychev, A.K.; Ivanov, A.; Lagarkov, A.; Barbillon, G. Light concentration by metal-dielectric micro-resonators for SERS sensing. *Materials* **2019**, *12*, 103. [CrossRef] [PubMed]
69. Powell, J.A.; Venkatakrishnan, K.; Tan, B. Hybridized enhancement of the SERS detection of chemical and bio-marker molecules through Au nanosphere ornamentation of hybrid amorphous/crystalline Si nanoweb nanostructure biochip devices. *J. Mater. Chem. B* **2016**, *4*, 5713–5728. [CrossRef]
70. Chan, S.; Kwon, S.; Koo, T.W.; Lee, L.P.; Berlin, A.A. Surface-enhanced Raman scattering of small molecules from silver-coated silicon nanopores. *Adv. Mater.* **2003**, *15*, 1595–1598. [CrossRef]
71. Wu, Y.K.; Liu, K.; Li, X.F.; Pan, S. Integrate silver colloids with silicon nanowire arrays for surface-enhanced Raman scattering. *Nanotechnology* **2011**, *22*, 215701. [CrossRef] [PubMed]
72. Lu, R.; Sha, J.; Xia, W.W.; Fang, Y.J.; Gu, L.; Wang, Y.W. A 3D-SERS substrate with high stability: Silicon nanowire arrays decorated by silver nanoparticles. *CrystEngComm* **2013**, *15*, 6207–6212. [CrossRef]
73. Wang, S.-Y.; Jiang, X.-X.; Xu, T.-T.; Wei, X.-P.; Lee, S.-T.; He, Y. Reactive ion etching-assisted surface-enhanced Raman scattering measurements on the single nanoparticle level. *Appl. Phys. Lett.* **2014**, *104*, 243104. [CrossRef]
74. Ding, S.-Y.; You, E.-M.; Tian, Z.-Q.; Moskovits, M. Electromagnetic theories of surface-enhanced Raman spectroscopy. *Chem. Soc. Rev.* **2017**, *46*, 4042–4076. [CrossRef] [PubMed]
75. Li, J.-F.; Zhang, Y.-J.; Ding, S.-Y.; Panneerselvam, R.; Tian, Z.-Q. Core–shell nanoparticle-enhanced Raman spectroscopy. *Chem. Rev.* **2017**, *117*, 5002–5069. [CrossRef]
76. Ding, S.-Y.; Yi, J.; Li, J.-F.; Ren, B.; Wu, D.-Y.; Panneerselvam, R.; Tian, Z.-Q. Nanostructure-based plasmon-enhanced Raman spectroscopy for surface analysis of materials. *Nat. Rev. Mater.* **2016**, *1*, 16021. [CrossRef]
77. Cao, Y.C.; Jin, R.; Mirkin, C.A. Nanoparticles with Raman spectroscopic fingerprints for DNA and RNA detection. *Science* **2002**, *297*, 1536–1540. [CrossRef]
78. Erdélyi, R.; Nagata, T.; Rogers, D.J.; Teherani, F.H.; Horváth, Z.E.; Lábadi, Z.; Baji, Z.; Wakayama, Y.; Volk, J. Investigations into the impact of the template layer on ZnO nanowire arrays made using low temperature wet chemical growth. *Cryst. Growth Des.* **2011**, *11*, 2515–2519. [CrossRef]
79. Özgür, Ü.; Alivov, Y.I.; Liu, C.; Teke, A.; Reshchikov, M.A.; Doğan, S.; Avrutin, V.; Cho, S.-J.; Morkoç, H. A comprehensive review of ZnO materials and devices. *J. Appl. Phys.* **2005**, *98*, 041301. [CrossRef]
80. Janotti, A.; Van de Walle, C.G. Fundamentals of zinc oxide as a semiconductor. *Rep. Prog. Phys.* **2009**, *72*, 126501. [CrossRef]
81. Cheng, C.; Yan, B.; Wong, S.M.; Li, X.; Zhou, W.; Yu, T.; Shen, Z.; Yu, H.; Fan, H.J. Fabrication and SERS Performance of Silver-Nanoparticle-Decorated Si/ZnO Nanotrees in Ordered Arrays. *ACS Appl. Mater. Interfaces* **2010**, *7*, 1824–1828. [CrossRef] [PubMed]
82. Lee, Y.; Lee, J.; Lee, T.K.; Park, J.; Ha, M.; Kwak, S.K.; Ko, H. Particle-on-film gap plasmons on antireflective ZnO nanocone arrays for molecular-level surface-enhanced Raman scattering sensors. *ACS Appl. Mater. Interfaces* **2015**, *7*, 26421–26429. [CrossRef] [PubMed]
83. Fang, Y.; Wang, Y.; Wan, Y.; Wang, Z.; Sha, J. Detailed study on photoluminescence property and growth mechanism of ZnO nanowire arrays grown by thermal evaporation. *J. Phys. Chem. C* **2010**, *114*, 12469–12476. [CrossRef]
84. Zhu, G.; Zhou, Y.; Wang, S.; Yang, R.; Ding, Y.; Wang, X.; Bando, Y.; Wang, Z.L. Synthesis of vertically aligned ultra-long ZnO nanowires on heterogeneous substrates with catalyst at the root. *Nanotechnology* **2012**, *23*, 055604. [CrossRef]
85. Chen, L.; Luo, L.; Chen, Z.; Zhang, M.; Antonio Zapien, J.; Lee, C.S.; Lee, S.T. ZnO/Au composite nanoarrays as substrates for surface-enhanced Raman scattering detection. *J. Phys. Chem. C* **2010**, *114*, 93–100. [CrossRef]
86. Shan, Y.; Yang, Y.; Cao, Y.; Fu, C.; Huang, Z. Synthesis of wheatear-like ZnO nanoarrays decorated with Ag nanoparticles and its improved SERS performance through hydrogenation. *Nanotechnology* **2016**, *27*, 145502. [CrossRef]
87. Xu, J.-Q.; Duo, H.-H.; Zhang, Y.-G.; Zhang, X.-W.; Fang, W.; Liu, Y.-L.; Shen, A.-G.; Hu, J.-M.; Huang, W.-H. Photochemical synthesis of shape-controlled nanostructured gold on zinc oxide nanorods as photocatalytically renewable sensors. *Anal. Chem.* **2016**, *88*, 3789–3795. [CrossRef]

88. He, X.; Yue, C.; Zang, Y.; Yin, J.; Sun, S.; Li, J.; Kang, J. Multi-hotspot configuration on urchin-like Ag nanoparticle/ZnO hollow nanosphere arrays for highly sensitive SERS. *J. Mater. Chem. A* **2013**, *1*, 15010–15015.
89. Park, S.G.; Jeon, T.Y.; Jeon, H.C.; Kwon, J.D.; Mun, C.; Lee, M.; Cho, B.; Kim, C.S.; Song, M.; Kim, D.H. Fabrication of Au-decorated 3D ZnO nanostructures as recyclable SERS substrates. *IEEE Sens. J.* **2016**, *16*, 3382–3386. [CrossRef]
90. Song, W.; Ji, W.; Vantasin, S.; Tanabe, I.; Zhao, B.; Ozaki, Y. Fabrication of a highly sensitive surface-enhanced Raman scattering substrate for monitoring the catalytic degradation of organic pollutants. *J. Mater. Chem. A* **2015**, *3*, 13556–13562. [CrossRef]
91. Jayram, N.D.; Sonia, S.; Poongodi, S.; Suresh Kumar, P.; Masuda, Y.; Mangalaraj, D.; Ponpandian, N.; Viswanathan, C. Superhydrophobic Ag decorated ZnO nanostructured thin film as effective surface enhanced Raman scattering substrates. *Appl. Surf. Sci.* **2015**, *355*, 969–977. [CrossRef]
92. Xie, Y.; Yang, S.; Mao, Z.; Li, P.; Zhao, C.; Cohick, Z.; Huang, P.-H.; Huang, T.J. In situ Fabrication of 3D Ag@ZnO nanostructures for microfluidic surface-enhanced Raman scattering systems. *ACS Nano* **2014**, *8*, 12175–12184. [CrossRef] [PubMed]
93. Marcí, G.; Augugliaro, V.; López-Muñoz, M.J.; Martín, C.; Palmisano, L.; Rives, V.; Schiavello, M.; Tilley, R.J.D.; Venezia, A.M. Preparation characterization and photocatalytic activity of polycrystalline ZnO/TiO$_2$ systems. 2. Surface, bulk characterization, and 4-Nitrophenol photodegradation in Liquid–Solid regime. *J. Phys. Chem. B* **2001**, *105*, 1033–1040. [CrossRef]
94. Zhou, Q.; Wen, J.Z.; Zhao, P.; Anderson, W.A. Synthesis of vertically-aligned zinc oxide nanowires and their application as a photocatalyst. *Nanomaterials* **2017**, *7*, 9. [CrossRef]

© 2019 by the authors. Licensee MDPI, Basel, Switzerland. This article is an open access article distributed under the terms and conditions of the Creative Commons Attribution (CC BY) license (http://creativecommons.org/licenses/by/4.0/).

Article

Gas Sensing with Nanoplasmonic Thin Films Composed of Nanoparticles (Au, Ag) Dispersed in a CuO Matrix

Manuela Proença *, Marco S. Rodrigues, Joel Borges * and Filipe Vaz

Centro de Física da Universidade do Minho, Campus de Gualtar, 4710-057 Braga, Portugal; marcopsr@gmail.com (M.S.R.); fvaz@fisica.uminho.pt (F.V.)
* Correspondence: manuelaproenca12@gmail.com (M.P.); joelborges@fisica.uminho.pt (J.B.); Tel.: +351-253-510-471 (J.B.)

Received: 6 May 2019; Accepted: 23 May 2019; Published: 25 May 2019

Abstract: Magnetron sputtered nanocomposite thin films composed of monometallic Au and Ag, and bimetallic Au-Ag nanoparticles, dispersed in a CuO matrix, were prepared, characterized, and tested, which aimed to find suitable nano-plasmonic platforms capable of detecting the presence of gas molecules. The Localized Surface Plasmon Resonance phenomenon, LSPR, induced by the morphological changes of the nanoparticles (size, shape, and distribution), and promoted by the thermal annealing of the films, was used to tailor the sensitivity to the gas molecules. Results showed that the monometallic films, Au:CuO and Ag:CuO, present LSPR bands at ~719 and ~393 nm, respectively, while the bimetallic Au-Ag:CuO film has two LSPR bands, which suggests the presence of two noble metal phases. Through transmittance-LSPR measurements, the bimetallic films revealed to have the highest sensitivity to the refractive index changes, as well as high signal-to-noise ratios, respond consistently to the presence of a test gas.

Keywords: thin films; magnetron sputtering; microstructure; noble metal nanoparticles; CuO matrix; localized surface plasmon resonance; gas sensor

---

1. Introduction

Nanocomposite thin films, containing noble metal nanoparticles embedded in an oxide matrix, have been a subject of considerable interest for optical gas sensing due to their localized surface plasmon resonance (LSPR) properties [1,2]. Surface plasmons are coherent oscillations of free electrons excited by an electromagnetic field at the boundaries between a metal and a dielectric. They can propagate along the surface of the conductor, which are designated by surface plasmon polaritons, or be confined to metallic nanoparticles or nanostructures, in which case, are denominated as localized surface plasmons [3–5]. LSPR can give rise to strong absorption bands, the enhancement of the electromagnetic field near the nanoparticles, and the appearance of scattering to the far field [6–10]. Since its discovery, there have been significant advances in both theoretical and experimental investigations of surface plasmons, which led to the development of new modelling methods that contribute to the understanding of the morphology and to the calculation of the optical properties of nano-plasmonic systems [8,11,12].

The two most well studied plasmonic metals are gold (Au) and silver (Ag). They exhibit LSPR bands within the visible spectrum due to the energy levels of d-d transitions, being used in various applications involving color [13,14] as well as in sensing due to their relatively high refractive index sensitivity [15,16]. Since Ag nanoparticles present the sharpest and strongest bands among all metals, they are associated to higher sensitivity factors than Au. However, Au nanoparticles are more frequently selected for sensing applications due to their lower toxicity, inert nature (less prone to oxidation), and stability [17,18]. On the other hand, Ag-Au bimetallic nanoparticles have attracted particular attention due to their corresponding monometallic counterparts, which may allow further improvements on their set of properties [19–21], especially the optical behavior [16,21,22]. In fact, they are relatively easy to prepare since both metals have a face-centred cubic structure and similar lattice constants. However, it is known that the synthesis method can result in alloyed bimetallic nanoparticles [23,24], core-shell [25], and even Janus systems [26]. These features are determined by the Au/Ag ratio in the bimetallic nanostructure, which leads to different optical properties. From the alloy formation of Au-Ag bimetallic nanoparticles, only one LSPR band results between the peaks of the constituting monometallic nanoparticles, while a mixed system originates two plasmonic bands, as reported in different works [3,22,23,27].

Diverse noble metal compositions dispersed in a dielectric matrix and different microstructures and nanostructures might be developed, which originates different LSPR bands, since their curvature and position are strongly dependent on different factors such as the composition, size, shape, and distribution of the nanoparticles, which are also sensitive to changes of the refractive index of the surrounding dielectric medium where they are dispersed [1,3]. Hence, the basis of the plasmonic bio/chemical sensors is established by the dependence of the LSPR band on the surrounding refractive index [2,16,28]. One of the advantages of using LSPR phenomenon for optical gas sensors in contrast to Surface Plasmon Resonance (SPR) systems is the fact that the first ones have a much higher potential to be sensitive to the extremely low refractive index changes such as those induced by gas molecules [2,29,30], since the plasmon decay length in LSPR is much lower than in SPR [31]. Furthermore, LSPR-based sensors are basically supported by nanoparticles that can be directly coupled to light, while the SPR-based sensors are dependent on prisms, optical fibers, or gratings to be coupled with light [30,32,33].

For the LSPR gas detection by refractive index changes to be functional, the production of highly sensitive plasmonic thin films is required, but the development of a high-resolution spectroscopy system to measure extremely small LSPR peak shifts is a fact that has been hampering the research on this area [2,34]. Hence, in order to optimize the sensitivity of the films, previous studies of the LSPR sensing response have been made by using two liquids with a relatively large refractive index difference [28,35,36], which allowed us to estimate the refractive index sensitivity (RIS) [37–39].

The present work proposes a reliable and effective possibility of sensitive thin films, suitable to be used as optical sensors. Such (nanocomposite) thin films are based on Au and/or Ag nanoparticles, dispersed in a semiconductor copper oxide (CuO) matrix, Au:CuO, Ag:CuO, and Au-Ag:CuO, deposited by reactive DC magnetron sputtering. The use of a pure copper target containing gold and/or silver pellets on its surface, avoids the use of a second cathode [40,41], with evident economic advantages [3,28]. After the preparation of the thin films, a thermal annealing treatment was performed in order to promote the necessary nanostructural changes in the noble metal nanoparticles, and dielectric matrix, which enabled the manifestation of the LSPR behavior, and, consequently, turned the thin films sensitive to the gas molecules. The composition and morphology of the thin films were studied and correlated with the LSPR responses. LSPR sensing tests were performed through transmittance measurements in a custom-made optical vacuum system, which incorporates a gas flow cell. The sensitivity of the different films to the presence of O_2 gas was also calculated and compared between them.

2. Materials and Methods

Thin films of Au:CuO, Ag:CuO, and Au-Ag:CuO manifesting LSPR behavior were produced by a two-step process, involving deposition of the thin films and posterior thermal treatment. For the depositions, two different types of substrates were used including Si (Boron doped, p type, <100> orientation, 525 µm thick) for chemical and (micro)structural characterization purposes and SiO_2 (fused silica) for optical spectra measurements. Before the depositions and in order to clean and activate the surface of the substrates, plasma treatments were performed by a Low-Pressure Plasma Cleaner by Diener Electronic equipped with a 40 kHz RF generator (Zepto Model, Ebhausen, Germany) [42], applying a power of 100 W. The substrates were first cleaned with O_2 plasma (80 Pa, for 5 min), and then activated with Ar plasma (80 Pa, for 15 min).

The films were deposited by reactive (DC) magnetron sputtering during 60 s in order to produce films with thicknesses around ~50 nm. As illustrated in Figure 1a, the above-mentioned substrates were then placed in a grounded hexagonal holder, rotating at 16 rpm and 7 cm far from the cathode. The latter is a rectangular copper target (200 × 100 × 6 mm^3, 99.99% purity), where gold and/or silver pellets (surface area of 960 mm^2 and 0.5 mm thick) were placed symmetrically on its preferential erosion zone. The base pressure was below 5×10^{-4} Pa, while the target potential was limited to 500 V, and the applied current was 3.25 mA/cm^2. The discharge was ignited in a gas atmosphere composed of Ar (3.5×10^{-1} Pa) and O_2 (2×10^{-2} Pa). Then, in order to promote the nanoparticles' growth, the films were subjected to thermal treatments in-air, up to a maximum temperature of 700 °C, according to what was previously studied and published by the group [1,28]. The heating ramp used was 5 °C/min and the isothermal period was 5 h, which cooled down freely inside the furnace, before reaching room temperature.

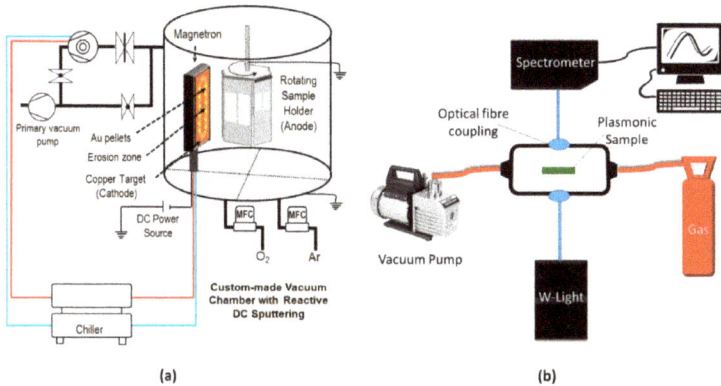

Figure 1. Simplified representation of the reactive DC magnetron sputtering system (**a**) and the custom-made system for transmittance-LSPR (T-LSPR) measurement in a controlled atmosphere (**b**).

The atomic composition of the films was studied by Rutherford Backscattering Spectrometry (RBS) using a Van de Graaff accelerator, a standard detector, placed at 140°, and two pin-diode detectors located symmetrically to each other, both placed at a 165° scattering angle respective to the beam direction. Spectra were collected using 2.0 MeV 4He$^+$, and 1.45 MeV 1H$^+$ beams at normal incidence and the data was analyzed with the IBA DataFurnace NDF v9.6i code [43].

The morphology of the films' surface was studied by a Dual Beam Scanning Electron Microscope, SEM/FIB FEI Helios 600i (Hillsboro, OR, USA), using a backscatter electron detector. The surface micrographs were analyzed using MATLAB software (version R2018a), by calculating the Feret diameter, the aspect ratio, and the nearest neighbor of the contrasted nanoparticles. The MATLAB algorithm included the locally adaptive threshold function "adaptthresh." After the binarization and scaling of the SEM images, the nanoparticles were analysed using the "regionprops" and "bwboundaries" functions.

The films' gas sensitivity was investigated by monitoring the LSPR band in the presence of O_2 (atmospheric pressure), in comparison to a low vacuum pressure. Real-time measurements were performed in a custom-made system (Figure 1b), composed of two main parts: the optical components and a vacuum system. The optical system allows the measurement of the optical (transmittance) spectrum of the sample, using a tungsten lamp and a modular spectrometer by Ocean Optics (HR4000, Edinburgh, UK). Optical fibers were used to connect those components to the flow cell, where the sample is placed. A vacuum pump was used to produce a "primary" vacuum (~40 Pa) inside the flow cell and then O_2 was introduced at atmospheric pressure for 120 s. Several vacuum/O_2 cycles were employed and the LSPR peak position was monitored in real time. A MATLAB algorithm was written to smooth the spectra and find the position of the LSPR peak over time.

3. Results

3.1. Thin Films Characterization

The atomic concentration profiles of the thin films were determined by RBS (Figure 2). The as-deposited CuO matrix (solid lines), and the CuO matrix with thermal treatment at 700 °C (dash lines), are represented in Figure 2a, while the as-deposited nanocomposite films are displayed in Figure 2b–d. According to the RBS analysis, all the as-deposited thin films were found to have a roughly constant atomic concentration across their thickness, even after the annealing process for the case of the pure matrix. Moreover, elemental concentration results revealed that the matrix of the as-deposited films is not fully CuO stoichiometric, since the atomic ratio C_O/C_{Cu} is always different from but close to 1. However, as soon as the film is subjected to thermal annealing, it seems that the CuO matrix becomes stoichiometric, which can be observed by the corresponding RBS profile (Figure 2a), where Cu and O concentrations were estimated to be about 50.0 ± 0.5 at.% and 50 ± 3 at.%, respectively. Thus, when the films are subjected to thermal treatment in air, the chemical composition may change in relation to the as-deposited films due to oxygen incorporation [44,45], as previously verified [1,28]. The atomic concentration of noble metals into the CuO matrix was determined to be about C_{Au} = 15.0 ± 0.5 at.% (Au:CuO), C_{Ag} = 17.7 ± 0.5 at.% (Ag:CuO), and C_{Au} = 6.7 ± 0.5 at.%, C_{Ag} = 8.0 ± 0.5 at.% (Au-Ag:CuO). These were the compositions of the thin films used for LSPR sensitivity tests.

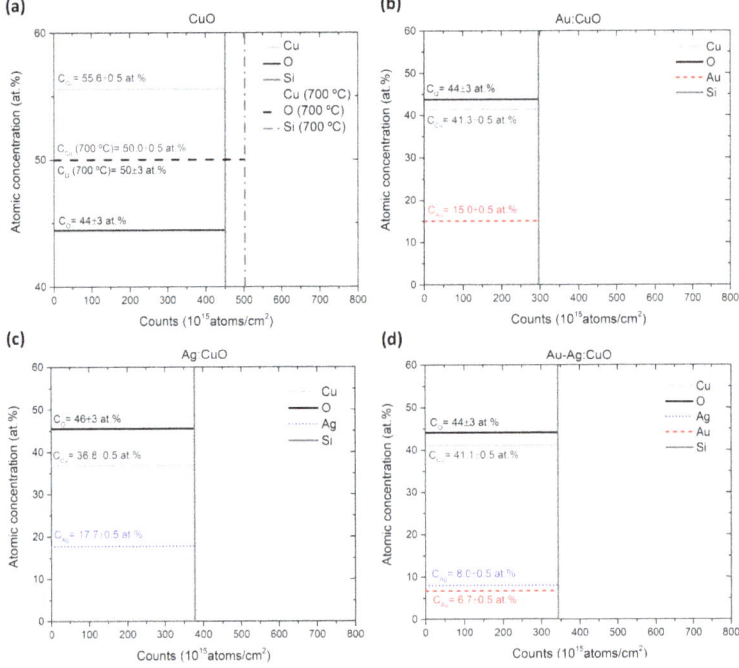

Figure 2. Atomic concentration (at.%) of the different elements present in the as-deposited CuO matrix (solid lines (**a**)), in the CuO matrix with annealing at 700 °C (dash lines (**a**)), and in the as-deposited samples of Au:CuO (**b**), Ag:CuO (**c**), and Au-Ag:CuO (**d**) films deposited with a pellets' area of 960 mm², obtained by the RBS data analyzed with the code IBA DataFurnace NDF v9.6i [43].

The CuO matrix annealed at 700 °C presents a polycrystalline structure with well-defined grain boundaries, as observed in the SEM micrograph displayed in Figure 3a. In addition, the optical transmittance spectrum (Figure 3b) reveals a semi-transparent CuO matrix in the visible range, with a progressive increase of transmittance for higher wavelengths, which is a feature that is in agreement with the literature [46].

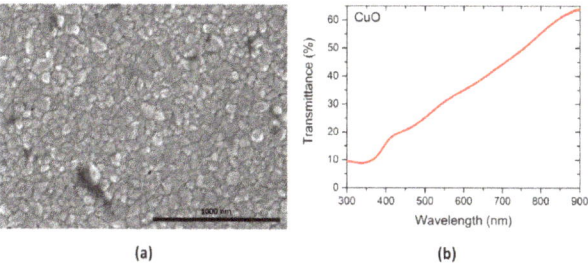

Figure 3. Top-view SEM micrograph of the CuO matrix annealed at 700 °C (**a**) and the respective optical transmittance spectrum (**b**).

The microstructural analysis of the annealed plasmonic thin films revealed the presence of noble nanoparticles (bright spots) in the different nanocomposite thin films ((a) and (b) in Figures 4–6), which suggests that the growth of nanoparticles might be facilitated by easier diffusion of Au and Ag atoms through grain boundaries of the CuO matrix. The Au:CuO (Figure 4) film is the one that presents the highest nanoparticles' density at the surface (127 µm^{-2}) with an average size of about 33 nm (Figure 4c). Moreover, the Au nanoparticles are relatively close to each other (Figure 4d) and they are presumably spherical since their aspect ratio distribution is narrow and close to 1, as seen in Figure 4e.

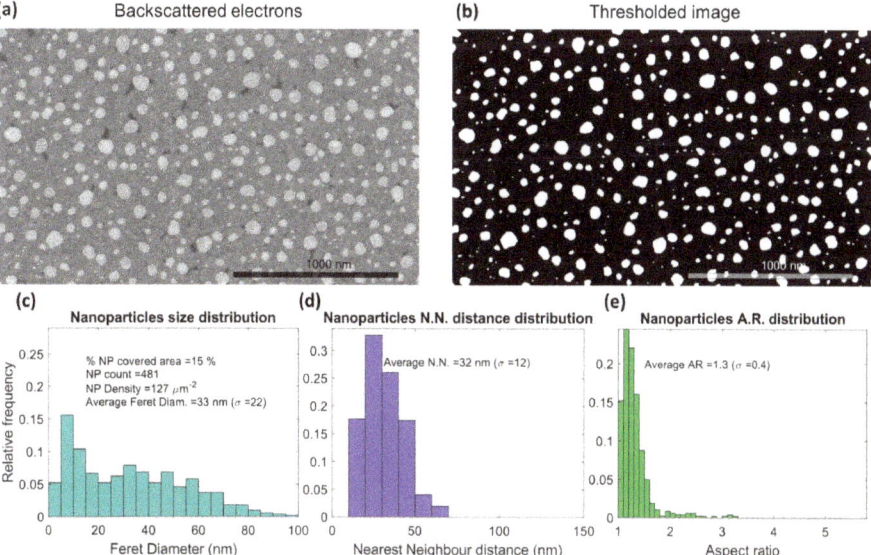

Figure 4. Au nanoparticle distribution analysis, using a MATLAB algorithm: (**a**) top-view SEM micrograph with atomic weight contrast and (**b**) processed and threshold micrograph showing the exposed Au nanoparticles, with 15% Au coverage area. Distribution histograms are displayed in (**c**) for the nanoparticles Feret diameter, (**d**) for the nearest neighbor distance, and (**e**) for the aspect ratio.

Regarding the Ag:CuO film (Figure 5), the average size of Ag nanoparticles was estimated to be 15 nm. However, the nanoparticles' density at the surface (69 µm²) is much smaller than in the other films (Figure 5c), which leads to the highest distance between the nanoparticles (Figure 5d). In fact, the formation of islands of Ag (micro-sized agglomerates with parallelepiped shape) was observed on the surface of the film (not shown here) [28]. This explains the low amount of nanosized Ag particles, which is a behavior that was not expected when taking into account the relatively high Ag atomic concentration determined for the as-deposited film.

The Au-Ag:CuO film (Figure 6) presents values between those belonging to the monometallic counterparts (Figures 4 and 5). It presents a density of Au-Ag nanoparticles at the surface of 100 µm^{-2}, with an average size estimated to be 30 nm (Figure 6c). Moreover, the nearest neighbor distance distribution is broader than in the Au film and narrower than in the Ag film. Moreover, this system shows the widest aspect ratio distribution, with an average value of 1.5, which proves that both spherical and irregular nanoparticles are present in the film's surface.

The different microstructures achieved by the films with the thermal treatment originated different optical transmittance responses, as shown in Figure 7. The high Au nanoparticles' density at the surface and their quasi-spherical shape, observed in the Au:CuO film (Figure 4), gave rise to a well-defined and sharp transmittance LSPR (T-LSPR) band at ~719 nm (Figure 7a), with a high transmittance amplitude, at about 15 percentage points (i.e., the difference between the maximum and the minimum band's peak).

On the other hand, a T-LSPR band was also observed for the Ag:CuO film (Figure 7b), appearing at shorter wavelengths (~393 nm) as is typical of the Ag nanoparticles [3,28]. However, despite the narrow shape, due to its slightly larger nanoparticle aspect ratio distribution, the LSPR band is also less intense since the number of Ag nanoparticles at the surface is scarce, which presents only a transmittance amplitude of ~10 percentage points.

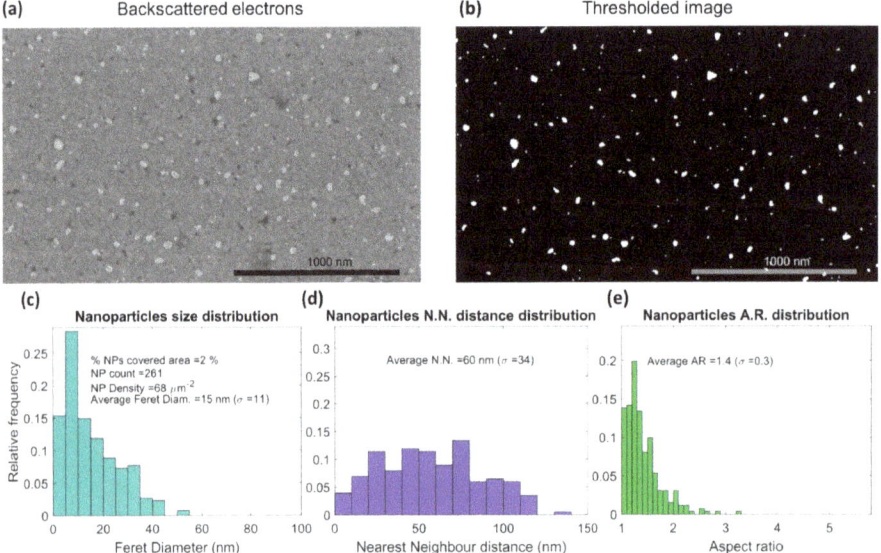

Figure 5. Ag nanoparticle distribution analysis, using a MATLAB algorithm: (**a**) top-view SEM micrograph with atomic weight contrast and (**b**) processed and threshold micrograph showing the exposed Ag nanoparticles, with 2% Ag coverage area. Distribution histograms are displayed in (**c**) for nanoparticles Feret diameter, (**d**) for the nearest neighbor distance, and (**e**) for the aspect ratio.

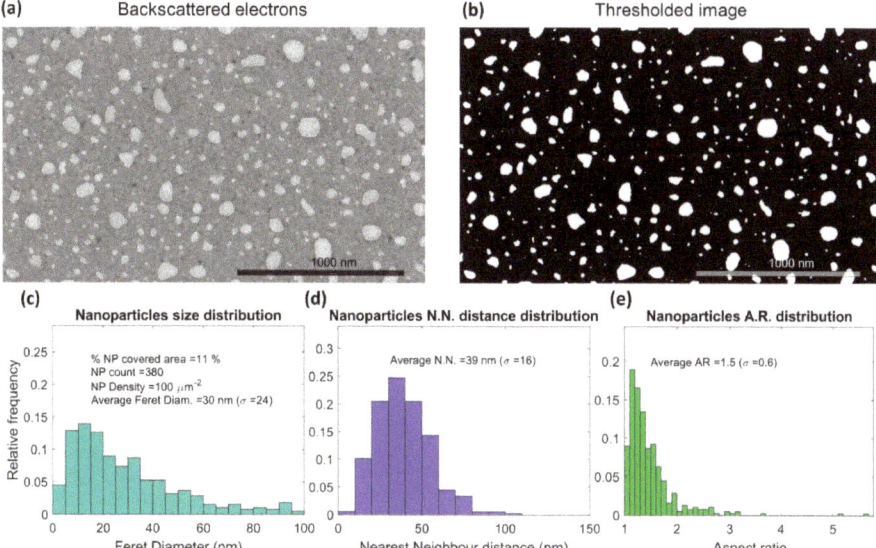

Figure 6. Au-Ag nanoparticle distribution analysis, using a MATLAB algorithm: (**a**) top-view SEM micrograph with atomic weight contrast and (**b**) processed and threshold micrograph showing the exposed Au-Ag nanoparticles, with 11% Au-Ag coverage area. Distribution histograms are displayed in (**c**) for nanoparticles Feret diameter, (**d**) for the nearest neighbor distance, and (**e**) for the aspect ratio.

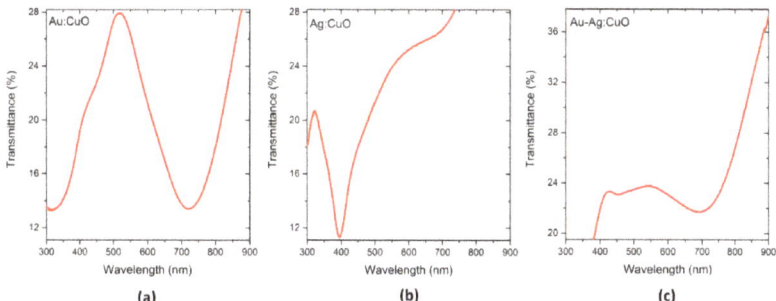

Figure 7. Transmittance spectra of the Au:CuO (**a**), Ag:CuO (**b**), and Au-Ag:CuO (**c**) thin films after in-air annealing.

Concerning the Au-Ag:CuO film, two shifted LSPR peaks are observed (~450 and 676 nm), even though the second one is much more pronounced (Figure 7c). The presence of two peaks might suggest the presence of separate phases of Ag and Au nanoparticles in these films, but since they are shifted from their initial positions, the formation of an alloy of Au-Ag bimetallic nanoparticles cannot be disregarded. Furthermore, as observed in Figure 6, the film presents both spherical and elongated nanoparticles, which contribute to the LSPR band widening and, therefore, appears much less intense.

3.2. Sensitivity Tests Using Exposure to O_2

In order to test the films' sensitivity to refractive index changes promoted by the presence and/or adsorption of gas molecules, they were exposed to a test gas (O_2). Figure 8 presents the LSPR peak position (transmittance) of the three systems, during five cycles under vacuum, and O_2 at atmospheric pressure. As expected from this type of sensor, the transmittance shift due to a change in the refractive index is typically very short, in the order of tenths of percentage points [2,47]. Anyway, it is possible to observe that the films responded consistently to the presence of the gas. The T-LSPR peak shifted to lower transmittances when the O_2 was introduced, which decreases by 0.35, 0.11, and 0.43 percentage points for the Au, Ag, and Au-Ag:CuO films, respectively. These results are consistent with what has been already published for Au-TiO$_2$ films, but with slightly higher sensitivities [2]. The Ag:CuO sample presents the lowest shift and, subsequently, the lowest signal-to-noise ratio (~3). This is believed to result from the morphology achieved after the annealing process (Figure 5), where the presence of Ag nanoparticles at the film's surface is scarce, which might hinder the film's sensitivity. Moreover, the presence of Au in the Au:CuO film and both Ag and Au nanoparticles in Au-Ag:CuO film seems to improve the film's response since a higher transmittance shift is observed when the test gas is introduced. In addition to show the highest transmittance shift, the Au-Ag:CuO film also presents the best signal-to-noise ratio (~123) even though the Au:CuO film has also a reasonable value of ~59. Furthermore, the peak shifts are reproducible every cycle when the test gas is introduced, which suggests that the eventual gas adsorption is reversible.

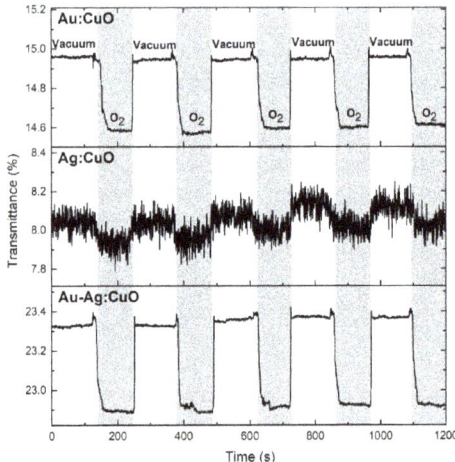

Figure 8. Variation of the LSPR peak position (transmittance minimum) of the Au:CuO, Ag:CuO, and Au-Ag:CuO films over time for five cycles of vacuum and O_2 atmosphere.

4. Conclusions

Au:CuO, Ag:CuO, and Au-Ag:CuO thin films with nanoplasmonic properties were produced in this work. The films were first deposited by magnetron sputtering for 1 min, using a Cu target with small metallic pellets (960 mm^2 pellet area) and a target potential limited to 500 V. Then, the different films were annealed up to 700 °C in order to promote the nanoparticles' growth and structural changes.

The composition analysis revealed the presence of reasonable amounts of noble metals in a CuO matrix, which becomes stoichiometric after a thermal treatment in air. Furthermore, the annealing induced structural and morphological changes that influenced the LSPR responses of the thin films. Due to the presence of spherical Au nanoparticles with high density at the surface, the Au:CuO film presented the most well-defined and pronounced transmittance LSPR band at ~719 nm, while the Ag:CuO film showed a narrower but less intense band at shorter wavelengths (~393 nm) due to the scarce number of Ag nanoparticles at the surface. However, despite the fact that the Au-Ag:CuO film has two T-LSPR peaks (~450 and 676 nm) with relatively low intensity, it showed to be the most sensitive to the refractive index changes, such as to the O_2 gas presence, followed after by Au:CuO and Ag:CuO films.

In conclusion, this work proves that the sensitivity of Au-Ag:CuO thin films to the test gas (O_2) can be improved by preparing bimetallic noble nanoparticles embedded in the CuO matrix. Hence, this configuration might be preferable to use for LSPR gas sensing.

Author Contributions: Conceptualization, J.B., F.V.; methodology, M.S.R., J.B.; software, M.S.R.; validation, M.P., M.S.R., J.B., F.V.; formal analysis, M.P., M.S.R., J.B., F.V.; investigation, M.P., M.S.R., J.B.; resources, J.B., F.V.; data curation, J.B., M.S.R., F.V.; writing—original Draft preparation, M.P., J.B.; writing—review and Editing, J.B., F.V.; visualization, M.P., M.S.R., J.B., F.V.; supervision, J.B., F.V.; project administration, F.V.; funding acquisition, J.B., F.V.

Funding: This research was funded by the Portuguese Foundation for Science and Technology (FCT) in the framework of the Strategic Funding UID/FIS/04650/2019; and by the projects NANOSENSING POCI-01-0145-FEDER-016902, with FCT reference PTDC/FIS-NAN/1154/2014; and project NANO4BIO POCI-01-0145-FEDER-032299, with FCT reference PTDC/FIS-MAC/32299/2017 supported this work. Manuela Proença acknowledges her PhD Scholarship from FCT, with reference SFRH/BD/137076/2018. Joel Borges acknowledges FCT for his Researcher Contract from project NANO4BIO, CTTI-149/18-CF(1). Marco S. Rodrigues acknowledges FCT for his PhD Scholarship, SFRH/BD/118684/2016.

Conflicts of Interest: The authors declare no conflict of interest.

References

1. Proença, M.; Borges, J.; Rodrigues, M.S.; Domingues, R.P.; Dias, J.P.; Trigueiro, J.; Bundaleski, N.; Teodoro, O.M.N.D.; Vaz, F. Development of Au/CuO nanoplasmonic thin films for sensing applications. *Surf. Coat. Technol.* **2018**, *343*, 178–185. [CrossRef]
2. Rodrigues, M.S.; Borges, J.; Proença, M.; Pedrosa, P.; Martin, N.; Romanyuk, K.; Kholkin, A.L.; Vaz, F. Nanoplasmonic response of porous Au-TiO$_2$ thin films prepared by oblique angle deposition. *Nanotechnology* **2019**, *30*, 22. [CrossRef]
3. Borges, J.; Ferreira, C.G.; Fernandes, J.P.C.; Rodrigues, M.S.; Proença, M.; Apreutesei, M.; Alves, E.; Barradas, N.P.; Moura, C.; Vaz, F. Thin films of Ag-Au nanoparticles dispersed in TiO$_2$: Influence of composition and microstructure on the LSPR and SERS responses. *J. Phys. D. Appl. Phys.* **2018**, *51*, 205102. [CrossRef]
4. Borges, J.; Kubart, T.; Kumar, S.; Leifer, K.; Rodrigues, M.S.; Duarte, N.; Martins, B.; Dias, J.P.; Cavaleiro, A.; Vaz, F. Microstructural evolution of Au/TiO$_2$ nanocomposite films: The influence of Au concentration and thermal annealing. *Thin Solid Films* **2015**, *580*, 77–88. [CrossRef]
5. Borges, J.; Pereira, R.M.S.; Rodrigues, M.S.; Kubart, T.; Kumar, S.; Leifer, K.; Cavaleiro, A.; Polcar, T.; Vasilevskiy, M.I.; Vaz, F. Broadband optical absorption caused by the plasmonic response of coalesced Au nanoparticles embedded in a TiO$_2$ matrix. *J. Phys. Chem. C* **2016**, *120*, 16931–16945. [CrossRef]
6. Ghosh, S.K.; Pal, T. Interparticle coupling effect on the surface plasmon resonance of gold nanoparticles: From theory to applications. *Chem. Rev.* **2007**, *107*, 4797–4862. [CrossRef]
7. Hutter, E.; Fendler, J.H. Exploitation of localized surface plasmon resonance. *Adv. Mater.* **2004**, *16*, 1685–1706. [CrossRef]
8. Toudert, J.; Simonot, L.; Camelio, S.; Babonneau, D. Advanced optical effective medium modeling for a single layer of polydisperse ellipsoidal nanoparticles embedded in a homogeneous dielectric medium: Surface plasmon resonances. *Phys. Rev. B* **2012**, *86*, 045415. [CrossRef]
9. Politano, A.; Formoso, V.; Chiarello, G. Dispersion and damping of gold surface plasmon. *Plasmonics* **2008**, *3*, 165–170. [CrossRef]
10. Pitarke, J.M.; Silkin, V.M.; Chulkov, E.V.; Echenique, P.M. Theory of surface plasmons and surface-plasmon polaritons. *Rep. Prog. Phys.* **2007**, *70*, 1. [CrossRef]
11. Scholl, J.A.; Koh, A.L.; Dionne, J.A. Quantum plasmon resonances of individual metallic nanoparticles. *Nature* **2012**, *483*, 421–427. [CrossRef]
12. Goyenola, C.; Gueorguiev, G.K.; Stafström, S.; Hultman, L. Fullerene-like CS$_x$: A first-principles study of synthetic growth. *Chem. Phys. Lett.* **2011**, *506*, 86–91. [CrossRef]
13. Rodrigues, M.S.; Borges, J.; Gabor, C.; Munteanu, D.; Apreutesei, M.; Steyer, P.; Lopes, C.; Pedrosa, P.; Alves, E.; Barradas, N.P.; et al. Functional behaviour of TiO$_2$ films doped with noble metals. *Surf. Eng.* **2016**, *32*, 554–561. [CrossRef]

14. Torrell, M.; Cunha, L.; Cavaleiro, A.; Alves, E.; Barradas, N.P.; Vaz, F. Functional and optical properties of Au:TiO$_2$ nanocomposite films: The influence of thermal annealing. *Appl. Surf. Sci.* **2010**, *256*, 6536–6542. [CrossRef]
15. Zhao, Y.; Yang, Y.; Cui, L.; Zheng, F.; Song, Q. Electroactive Au@Ag nanoparticles driven electrochemical sensor for endogenous H$_2$S detection. *Biosens. Bioelectron.* **2018**, *117*, 53–59. [CrossRef] [PubMed]
16. Ghodselahi, T.; Arsalani, S.; Neishabooryneiad, T. Synthesis and biosensor application of Ag@Au bimetallic nanoparticles based on localized surface plasmon resonance. *Appl. Surf. Sci.* **2014**, *301*, 230–234. [CrossRef]
17. Borges, J.; Buljan, M.; Sancho-Parramon, J.; Bogdanovic-Radovic, I.; Siketic, Z.; Scherer, T.; Kübel, C.; Bernstorff, S.; Cavaleiro, A.; Vaz, F.; et al. Evolution of the surface plasmon resonance of Au:TiO$_2$ nanocomposite thin films with annealing temperature. *J. Nanopart. Res.* **2014**, *16*, 2790. [CrossRef]
18. Petryayeva, E.; Krull, U.J. Localized surface plasmon resonance: Nanostructures, bioassays and biosensing—A review. *Anal. Chim. Acta* **2011**, *706*, 8–24. [CrossRef] [PubMed]
19. Cesca, T.; Michieli, N.; Kalinic, B.; Balasa, I.G.; Rangel-Rojo, R.; Reyes-Esqueda, J.A.; Mattei, G. Bidimensional ordered plasmonic nanoarrays for nonlinear optics, nanophotonics and biosensing applications. *Mater. Sci. Semicond. Process.* **2019**, *92*, 2–9. [CrossRef]
20. Dwivedi, C.; Chaudhary, A.; Srinivasan, S.; Nandi, C.K. Polymer stabilized bimetallic alloy nanoparticles: Synthesis and catalytic application. *Colloid Interface Sci. Commun.* **2018**, *24*, 62–67. [CrossRef]
21. Khlebtsov, B.N.; Liu, Z.; Ye, J.; Khlebtsov, N.G. Au@Ag core/shell cuboids and dumbbells: Optical properties and SERS response. *J. Quant. Spectrosc. Radiat. Transf.* **2015**, *167*, 64–75. [CrossRef]
22. Sangpour, P.; Akhavan, O.; Moshfegh, A.Z. The effect of Au/Ag ratios on surface composition and optical properties of co-sputtered alloy nanoparticles in Au-Ag:SiO$_2$ thin films. *J. Alloy. Compd.* **2009**, *486*, 22–28. [CrossRef]
23. Sangpour, P.; Akhavan, O.; Moshfegh, A.Z. rf reactive co-sputtered Au-Ag alloy nanoparticles in SiO$_2$ thin films. *Appl. Surf. Sci.* **2007**, *253*, 7438–7442. [CrossRef]
24. Hareesh, K.; Joshi, R.P.; Sunitha, D.V.; Bhoraskar, V.N.; Dhole, S.D. Anchoring of Ag-Au alloy nanoparticles on reduced graphene oxide sheets for the reduction of 4-nitrophenol. *Appl. Surf. Sci.* **2016**, *389*, 1050–1055. [CrossRef]
25. Tiunov, I.A.; Gorbachevskyy, M.V.; Kopitsyn, D.S.; Kotelev, M.S.; Ivanov, E.V.; Vinokurov, V.A.; Novikov, A.A. Synthesis of large uniform gold and core–shell gold–silver nanoparticles: Effect of temperature control. *Russ. J. Phys. Chem. A* **2016**, *90*, 152–157. [CrossRef]
26. Song, Y.; Liu, K.; Chen, S. AgAu bimetallic janus nanoparticles and their electrocatalytic activity for oxygen reduction in alkaline media. *Langmuir* **2012**, *28*, 17143–17152. [CrossRef] [PubMed]
27. Blaber, M.G.; Arnold, M.D.; Harris, N.; Ford, M.J.; Cortie, M.B. Plasmon absorption in nanospheres: A comparison of sodium, potassium, aluminium, silver and gold. *Phys. B Condens. Matter* **2007**, *394*, 184–187. [CrossRef]
28. Proença, M.; Borges, J.; Rodrigues, M.S.; Meira, D.I.; Sampaio, P.; Dias, J.P.; Pedrosa, P.; Martin, N.; Bundaleski, N.; Teodoro, O.M.N.D.; et al. Nanocomposite thin films based on Au-Ag nanoparticles embedded in a CuO matrix for localized surface plasmon resonance sensing. *Appl. Surf. Sci.* **2019**, *484*, 152–168. [CrossRef]
29. Honda, M.; Ichikawa, Y.; Rozhin, A.G.; Kulinich, S.A. UV plasmonic device for sensing ethanol and acetone. *Appl. Phys. Express* **2018**, *11*, 012001. [CrossRef]
30. Kreno, L.E.; Hupp, J.T.; Van Duyne, R.P. Metal-organic framework thin film for enhanced localized surface Plasmon resonance gas sensing. *Anal. Chem.* **2010**, *82*, 8042–8046. [CrossRef]
31. Sagle, L.B.; Ruvuna, L.K.; Ruemmele, J.A.; Van Duyne, R.P. Advances in localized surface plasmon resonance spectroscopy biosensing. *Nanomedicine* **2011**, *6*, 1447–1462. [CrossRef] [PubMed]
32. Hammond, J.L.; Bhalla, N.; Rafiee, S.D.; Estrela, P. Localized surface plasmon resonance as a biosensing platform for developing countries. *Biosensors* **2014**, *4*, 172–188. [CrossRef] [PubMed]
33. Haes, A.J.; Zou, S.; Schatz, G.C.; Van Duyne, R.P. Nanoscale optical biosensor: Short range distance dependence of the localized surface plasmon resonance of noble metal nanoparticles. *J. Phys. Chem. B* **2004**, *108*, 6961–6968. [CrossRef]
34. Bingham, J.M.; Anker, J.N.; Kreno, L.E.; Duyne, R.P. Van gas sensing with high-resolution localized surface plasmon resonance spectroscopy. *J. Am. Chem. Soc.* **2010**, *132*, 17358–17359. [CrossRef]

35. Demirdjian, B.; Bedu, F.; Ranguis, A.; Ozerov, I.; Henry, C.R. Water adsorption by a sensitive calibrated gold plasmonic nanosensor. *Langmuir* **2018**, *34*, 5381–5385. [CrossRef]
36. Jeong, H.H.; Mark, A.G.; Alarcón-Correa, M.; Kim, I.; Oswald, P.; Lee, T.C.; Fischer, P. Dispersion and shape engineered plasmonic nanosensors. *Nat. Commun.* **2016**, *7*, 11331. [CrossRef] [PubMed]
37. Chen, P.; Liedberg, B. Curvature of the localized surface plasmon resonance peak. *Anal. Chem.* **2014**, *86*, 7399–7405. [CrossRef]
38. Kedem, O.; Vaskevich, A.; Rubinstein, I. Critical issues in localized plasmon sensing. *J. Phys. Chem. C* **2014**, *118*, 8227–8244. [CrossRef]
39. Jung, L.S.; Campbell, C.T.; Chinowsky, T.M.; Mar, M.N.; Yee, S.S. Quantitative interpretation of the response of surface plasmon resonance sensors to adsorbed films. *Langmuir* **1998**, *14*, 5636–5648. [CrossRef]
40. Das, S.; Alford, T.L. Structural and optical properties of Ag-doped copper oxide thin films on polyethylene napthalate substrate prepared by low temperature microwave annealing. *J. Appl. Phys.* **2013**, *113*, 244905. [CrossRef]
41. Rydosz, A.; Szkudlarek, A. Gas-sensing performance of M-doped CuO-based thin films working at different temperatures upon exposure to propane. *Sensors* **2015**, *15*, 20069–20085. [CrossRef]
42. Pedrosa, P.; Fiedler, P.; Lopes, C.; Alves, E.; Barradas, N.P.; Haueisen, J.; Machado, A.V.; Fonseca, C.; Vaz, F. Ag:TiN-coated polyurethane for dry biopotential electrodes: From polymer plasma interface activation to the first EEG measurements. *Plasma Process. Polym.* **2016**, *13*, 341–354. [CrossRef]
43. Barradas, N.P.; Jeynes, C. Advanced physics and algorithms in the IBA DataFurnace. *Nucl. Instrum. Methods Phys. Res. Sect. B* **2008**, *266*, 1875–1879. [CrossRef]
44. Liu, Y.; Zhang, J.; Zhang, W.; Liang, W.; Yu, B.; Xue, J. Effects of annealing temperature on the properties of copper films prepared by magnetron sputtering. *J. Wuhan Univ. Technol. Mater. Sci. Ed.* **2015**, *30*, 92–96. [CrossRef]
45. Pierson, J.F.; Wiederkehr, D.; Billard, A. Reactive magnetron sputtering of copper, silver, and gold. *Thin Solid Films* **2005**, *478*, 196–205. [CrossRef]
46. Figueiredo, V.; Elangovan, E.; Gonçalves, G.; Barquinha, P.; Pereira, L.; Franco, N.; Alves, E.; Martins, R.; Fortunato, E. Effect of post-annealing on the properties of copper oxide thin films obtained from the oxidation of evaporated metallic copper. *Appl. Surf. Sci.* **2008**, *254*, 3949–3954. [CrossRef]
47. Borensztein, Y.; Delannoy, L.; Djedidi, A.; Barrera, R.G.; Louis, C. Monitoring of the plasmon resonance of gold nanoparticles in Au/TiO$_2$ catalyst under oxidative and reducing atmospheres. *J. Phys. Chem. C* **2010**, *114*, 9008–9021. [CrossRef]

© 2019 by the authors. Licensee MDPI, Basel, Switzerland. This article is an open access article distributed under the terms and conditions of the Creative Commons Attribution (CC BY) license (http://creativecommons.org/licenses/by/4.0/).

Article

Preparation of Orthorhombic WO₃ Thin Films and Their Crystal Quality-Dependent Dye Photodegradation Ability

Yuan-Chang Liang [1],* and Che-Wei Chang [2]

[1] Institute of Materials Engineering, National Taiwan Ocean University, Keelung 20224, Taiwan
[2] Undergraduate Program in Optoelectronics and Materials Technology, National Taiwan Ocean University, Keelung 20224, Taiwan; jf860218@gmail.com
* Correspondence: yuanvictory@gmail.com

Received: 8 January 2019; Accepted: 1 February 2019; Published: 2 February 2019

Abstract: Direct current (DC) magnetron sputtering deposited WO_3 films with different crystalline qualities were synthesized by postannealing at various temperatures. The in-situ DC sputtering deposited WO_3 thin film at 375 °C exhibited an amorphous structure. The as-grown WO_3 films were crystallized after annealing at temperatures of 400–600 °C in ambient air. Structural analyses revealed that the crystalline WO_3 films have an orthorhombic structure. Moreover, the crystallite size of the WO_3 film exhibited an explosive coarsening behavior at an annealing temperature above 600 °C. The density of oxygen vacancy of the WO_3 films was substantially lowered through a high temperature annealing procedure. The optical bandgap values of the WO_3 films are highly associated with the degree of crystalline quality. The annealing-induced variation of microstructures, crystallinity, and bandgap of the amorphous WO_3 thin films explained the various photoactivated properties of the films in this study.

Keywords: sputtering; annealing; crystal quality; photoactivated properties

---

1. Introduction

Tungsten oxide (WO_3), as a wide bandgap semiconductor, has been intensively investigated for various uses in scientific devices [1,2]. Among various applications, the photocatalyst application for degrading organic pollutants receives much attention as WO_3 has the advantages of low cost, high chemical stability, and excellent process-dependent reproducibility. In general, pure WO_3 crystal shows five phase transitions at temperatures ranging from −180 to 900 °C [3]. Among the five crystal forms, a monoclinic I WO_3 phase is the stable phase at room temperature. However, the orthorhombic WO_3 phase exists only in some WO_3 nanostructures at room temperature, not frequently visible for other morphologies such as a thin-film structure. This is attributable to the fact that the transition temperature of the orthorhombic WO_3 phase for WO_3 nanostructures is generally at room temperature, which is quite lower than that of bulk WO_3.

WO_3 in a thin-film structure is highly desirable for various device applications because thin solid film can be integrated into various small devices or combined with other materials to form composites for scientific applications. Several methods of manufacturing WO_3 thin films with various microstructures for applications have been reported. For example, the pulsed-laser deposited WO_3 thin films are integrated with TiO_2 thin films to form multilayer films and used for photodegradation of methylene blue (MB) solution. The WO_3 layer in the multilayer structure enhances the photocatalytic ability of the TiO_2 layer [4]. The spray pyrolysis synthesized WO_3 thin films have also been used for photodegrading methyl orange (MO) [5]. The thermal evaporation deposited WO_3 thin films with adequate postannealing procedures in an oxygen-rich environment at 500 °C for 1 h have been used to degrade MB under irradiation [6]. Radio-frequency (RF) sputtering deposited WO_3/TiO_2 bilayer thin films with various WO_3 content are used to enhance the photocatalytic activity of TiO_2-based materials [7]. However, most WO_3 thin films synthesized through various methods are in a monoclinic structure; that is, the orthorhombic structure is limited in number.

Among various physical synthesis methods for WO_3 thin films, sputtering has been widely used to prepare oxide thin films with controllable microstructures and tunable physical properties. Moreover, the DC sputtering deposited WO_3 thin films via a metallic tungsten target provide the advantages of low cost and highly recycled usage of the target in comparison with those in WO_3 ceramic targeted by radio-frequency sputtering. Although DC sputtering growth of monoclinic or amorphous WO_3 thin films have been investigated for applications in gas-sensing, photocatalytic, and electrochromic devices, reports on the microstructure-dependent photodegradation properties of DC sputtering deposited orthorhombic WO_3 photocatalysts toward organic dyes are still lacking. Karuppasamy et al. synthesized amorphous WO_3 thin films via DC sputtering under various oxygen pressures. This work revealed that WO_3 films deposited at a lower working pressure exhibit satisfactory electrochromic properties. The variation of bulk density of the films prepared at various oxygen pressures affects the efficiency of insertion and removal of protons and electrons [8]. Kim et al. used DC sputtering to deposit monoclinic WO_3 films with various degrees of crystallinity by controlling the growth temperature from 200–500 °C. Higher crystal quality of WO_3 films deposited at a higher temperature resulted in better gas-sensing ability [9]. WO_3 films consisting of nanostructured surface feature are prepared by DC sputtering at various working pressures. The as-synthesized monoclinic WO_3 thin films are further used to photodegrade stearic acid [10]. Moreover, Stolze et al. prepared amorphous WO_3 films via DC sputtering under various O_2 percentages ranging from 0 to 20 vol.%; the effects of stoichiometry of WO_3 films on electrochromic properties were discussed [11]. The aforementioned examples demonstrate that most work on DC sputtering WO_3 films mainly focused on the monoclinic or amorphous phase. Reports on in-situ DC sputtering growth of orthorhombic WO_3 thin films are limited in number; this is associated with the fact that the orthorhombic structured WO_3 is unstable when the substrate temperature of the sputtering process is cooled down to room temperature. Therefore, the strategy to resolve this issue is to grow amorphous WO_3 films under a low oxygen content atmosphere, and then conduct various annealing procedures to obtain the stable orthorhombic WO_3 phase. Furthermore, the microstructure-dependent photocatalytic properties of the orthorhombic WO_3 films are presented in this study. The results herein might be useful for designing orthorhombic WO_3 thin-film photocatalysts with a desirable photodegradation ability toward organic dyes.

2. Methods

In this study, the WO_3 thin films were grown on 300 nm-thick SiO_2/Si and glass substrates by reactive DC magnetron sputtering at 375 °C. The metallic tungsten (purity > 99.99%) target was employed. The sputtering power of tungsten was kept at 40 W and the working pressure was maintained at 1.33 Pa during sputtering. The ratio of Ar/O_2 was fixed at 6:1. The thin film thickness was controlled to be 120 nm. The as-grown WO_3 films were subsequently annealed at 400–600 °C in ambient air for 1 h.

Scanning electron microscopy (SEM; S-4800, Hitachi, Tokyo, Japan) was used to investigate surface morphology of the WO_3 thin films. The surface roughness of various thin-film samples was measured by atomic focus microscopy (AFM; D5000, Veeco Karlsruhe, Germany). Crystal structures of the films were investigated by X-ray diffraction (XRD; D2 Phaser, Bruker, Karlsruhe, Germany). The detailed microstructures of the films were studied by high-resolution transmission electron microscopy (HRTEM; JEM-2100F, JEOL Tokyo, Japan). An X-ray photoelectron spectroscope (XPS; PHI 5000 VersaProbe, ULVAC-PHI, Chigasaki, Japan) was used to understand the chemical binding states of the thin films' elements. The transmittance spectra of the thin films were measured using a UV-Vis spectrophotometer (V750, Jasco, Tokyo, Japan). Comparison of photocatalytic activity of various thin films samples was performed using 10 mL methylene blue (MB; 10^{-6} M) solution containing various

WO₃ thin films under various irradiation conditions. The change of MB solution concentration after photodegradation tests was analyzed by measuring the intensity variation of absorbance spectra using an UV-Vis spectrophotometer.

3. Result and Discussions

The change in crystal structure features of the WO$_3$ thin films with various thermal annealing procedures is depicted in Figure 1. The as-grown WO$_3$ film shows an amorphous structure, as no visible Bragg reflections are found. Figure 1 shows the XRD patterns of the WO$_3$ films with thermal annealing procedures at 400–600 °C. The orthorhombic crystalline WO$_3$ phase is formed with distinguishable Bragg reflections (JCDPS. 20-1324). The more intense Bragg reflections associated with a narrower full-width at half maximum were observed for the thin films annealed at a higher annealing temperature, revealing a higher degree of crystalline quality of the film. Notably, no other peaks of impurities were observed after thermal annealing procedures. The XRD results reveal that the as-grown WO$_3$ thin films exhibited a polycrystalline feature after postannealing. The crystallite sizes of the annealed WO$_3$ films were evaluated using the Scherrer formula [12]. The crystallite sizes of the WO$_3$ films annealed at 400, 500, and 600 °C were approximately 24, 32, and 57 nm, respectively. The (001)-oriented crystal dominated the crystal structure feature of the crystalline WO$_3$ thin films annealed below 500 °C; moreover, the (111)-oriented crystal dominated the crystal structure feature when the film was annealed at 600 °C. Notably, the change of the WO$_3$ crystal orientation from (001) to (111) at the higher temperature annealing is associated with the surface binding energy among the low index crystallographic planes [13].

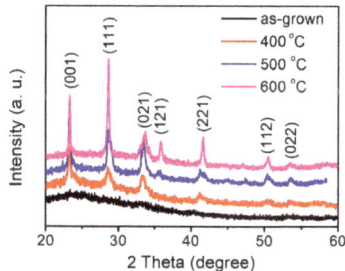

Figure 1. XRD patterns of the WO$_3$ films with and without thermal annealing procedures.

Surface morphologies of the WO$_3$ thin films with and without thermal annealing are shown in Figure 2. No distinctly well surface grain features can be seen for the as-grown WO$_3$ (Figure 2a). This might be associated with the amorphous nature of the sample as characterized by the XRD measurement. When the film was annealed at 400 °C, a surface grain feature was visible and the surface grains had an average size of approximately 32 nm (Figure 2b). Further increasing the annealing temperature to 500 °C increased the size of surface grains, and the homogeneity of grain size improved simultaneously. The average surface grain size was approximately 51 nm evaluated from Figure 2c. Notably, the surface grain size was abnormally increased (average grain size of 102 nm) and an uniformly cylindrical crystal feature was obtained when the film was annealed at 600 °C (Figure 2d). The high annealing temperature provides sufficient energy, which might facilitate the coalescence of the adjacent tiny crystals, and therefore large surface grains were formed.

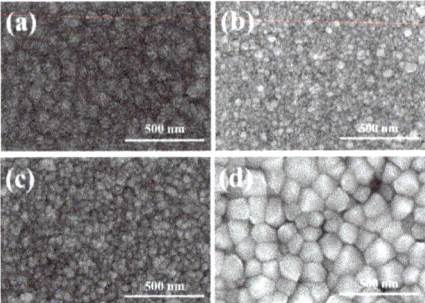

Figure 2. SEM images of the WO$_3$ thin films with and without annealing: (**a**) as-grown, (**b**) 400 °C, (**c**) 500 °C, (**d**) 600 °C.

Furthermore, the surface roughness of the various WO$_3$ thin films was further characterized by AFM. Figure 3a exhibits the surface of the as-grown WO$_3$ thin film. The root mean square (RMS) roughness of the as-grown amorphous WO$_3$ thin film was evaluated to be approximately 3.55 nm. Comparatively, the WO$_3$ thin films annealed at 400–600 °C exhibited coarser surface morphology (Figure 3b–d). The RMS roughness values of the WO$_3$ thin films were of approximately 4.02, 4.75, and 9.28 nm corresponding to the annealing temperature of 400, 500, and 600 °C, respectively. This result demonstrated that the surface roughness monotonically increases with increasing annealing temperature because high annealing temperature facilitates the coalescence of the surface grains and therefore rougher surface. The average surface grain sizes of the WO$_3$ thin films annealed at 400, 500, and 600 °C were approximately 26, 43, and 84 nm, respectively. Larger surface grains of the annealed film engendered a rougher surface feature. Similarly, a substantially increased surface grain size, as reported in the CuO film, annealed at the temperature higher than 700 °C [14].

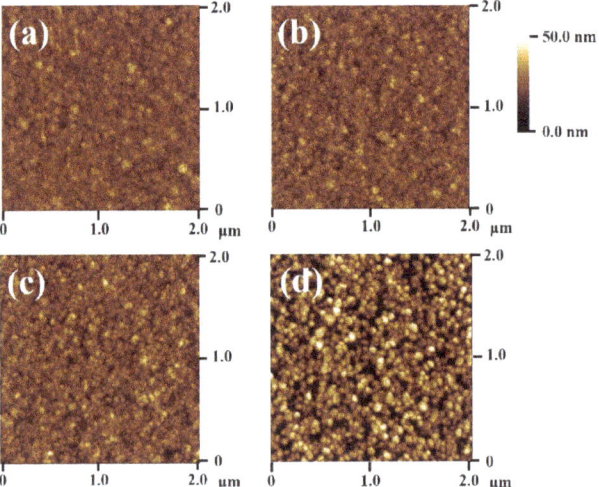

Figure 3. AFM images of WO$_3$ thin films with and without annealing: (**a**) as-grown, (**b**) 400 °C, (**c**) 500 °C, (**d**) 600 °C.

The detailed microstructures of the WO$_3$ thin films with and without thermal annealing at 600 °C were investigated by TEM. A low-magnification, cross-sectional TEM image of the as-grown WO$_3$ thin film is shown in Figure 4a. The thickness of the WO$_3$ film was ~120 nm. The film surface is dense and smooth, and no voids can be seen. A high-resolution TEM (HRTEM) micrograph of the as-grown WO$_3$ thin film is depicted in Figure 4b. The random and chaotic lattice fringes with a short-range order are distributed over the area of interest, revealing that the film is in the amorphous phase. Moreover, the selected area electron diffraction (SAED) pattern in Figure 4c exhibits a faint ring-like pattern, revealing that the film without heat treatment is uncrystallized. This is in agreement with the XRD result. Figure 4d depicts the energy-dispersive X-ray spectroscopy (EDS) spectra of the film, confirming that the film's composition consisted of W and O. Moreover, the O/W composition ratio is approximately 2.48, demonstrating oxygen deficiency in the WO$_3$ thin film. This is often observed in oxide thin films prepared by sputtering because the thin film growth condition is in an oxygen deficient environment during sputtering [15].

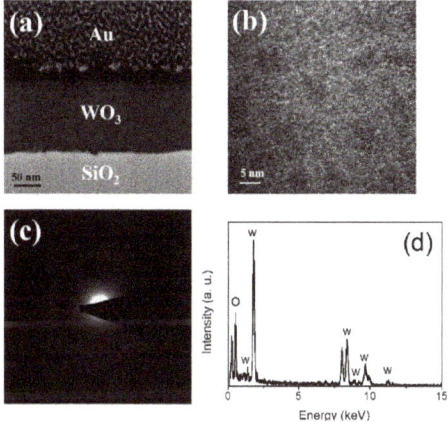

Figure 4. TEM analysis of the as-grown WO$_3$ film: (**a**) low-magnification image, (**b**) high-resolution image, (**c**) SAED pattern, (**d**) EDS spectra.

Figure 5a depicts a low-magnification, cross-sectional image of the WO$_3$ film annealed at 600 °C. The film thickness of the annealed WO$_3$ film is homogeneous throughout its cross section. Compared to the as-grown film, the surface and root of the high-temperature-annealed film are more undulated. Figure 5b,c demonstrate HRTEM images of the annealed WO$_3$ thin film. The appearance of visible and ordered lattice fringes in the HREM images indicate that the WO$_3$ film after annealing had a high degree of crystallinity. The atomic lattice fringes with intervals of approximately 0.39, 0.31, and 0.27 nm could be identified and were attributed to the interplanar distances of the WO$_3$ (001), (111), and (021) crystallographic planes, respectively. The boundaries between the adjacent grains were visible. The polycrystalline nature and the orthorhombic structure of the WO$_3$ film were also confirmed by the SAED measurements in Figure 5d. Distinct diffraction spots arranged in centric rings revealed the crystalline WO$_3$ thin film was formed after the 600 °C annealing process. Figure 5e shows the EDS spectra; the spectra revealed that the film mainly composed of W and O. No other impurity atoms were detected.

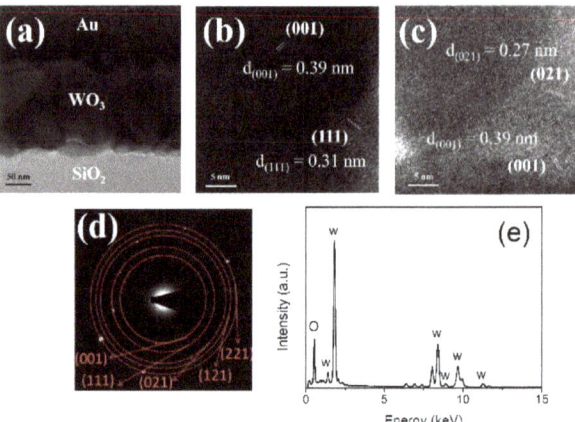

Figure 5. TEM analysis of the WO$_3$ film annealed at 600 °C: (**a**) low-magnification image, (**b**,**c**) HR images, (**d**) SAED pattern, (**e**) EDS spectra.

XPS analysis was performed to reveal the elemental binding states of various WO$_3$ thin films. The annealing temperature-dependent W oxidation state change is shown in Figure 6a–d. From the figures, the intense doublet with binding energies of approximately 35.0 eV (W4$f_{7/2}$) and 37.2 eV (W4$f_{5/2}$) are associated with photoelectrons emitted from W^{6+} ions of the WO$_3$ films, while the relatively small peaks at 34.0 and 36.2 eV can be assigned to W4$f_{7/2}$ and W4$f_{5/2}$ of W^{5+} oxidation state in tungsten oxides [16]. The presence of W^{5+} suggests the existence of crystal defects in the WO$_3$ film. Comparatively, the area and the height of core level W^{5+} decreased after annealing, which implied increased oxidation states of W in the WO$_3$ film. Notably, the WO$_3$ film annealed at 600 °C had the smallest features of W^{5+}, which indicates the surface tungsten in this film exhibited a larger degree of oxidation state after annealing. No peaks attributed to metallic W were identified in the spectra of all films. Notably, the W/O atomic ratio of the as-grown WO$_3$ film was approximately 0.4. Moreover, the W/O atomic ratio of the WO$_3$ films decreased from 0.37 to 0.34 with the annealing temperature increasing from 400 to 600 °C, respectively, evaluated from the XPS analyses. Figure 6e–h show that the XPS spectra of O1s for various WO$_3$ thin films have an asymmetric curve feature. The O1s spectra of the surface of various WO$_3$ thin films were fitted by two distributions, centered at approximately 529.3 and 530.8 eV, respectively. The relatively low binding energy peak is attributed to O^{2-} ions in the oxide lattice. The higher binding energy peak is attributed to the oxygen vacancies in the WO$_3$ [17]. The relative content of the oxygen vacancy for various WO$_3$ films was evaluated according to the area ratio of these two deconvolution components: (red peak)/(red peak + blue peak). The relative area of the higher energy binding component for the WO$_3$ films decreased with the annealing temperature. A great amount of vacancy existed in the surface of the as-grown WO$_3$ film. After annealing, the surface oxygen vacancy content markedly decreased from 34.4% to 25.3% with the annealing temperature increasing from 400 to 600 °C, respectively.

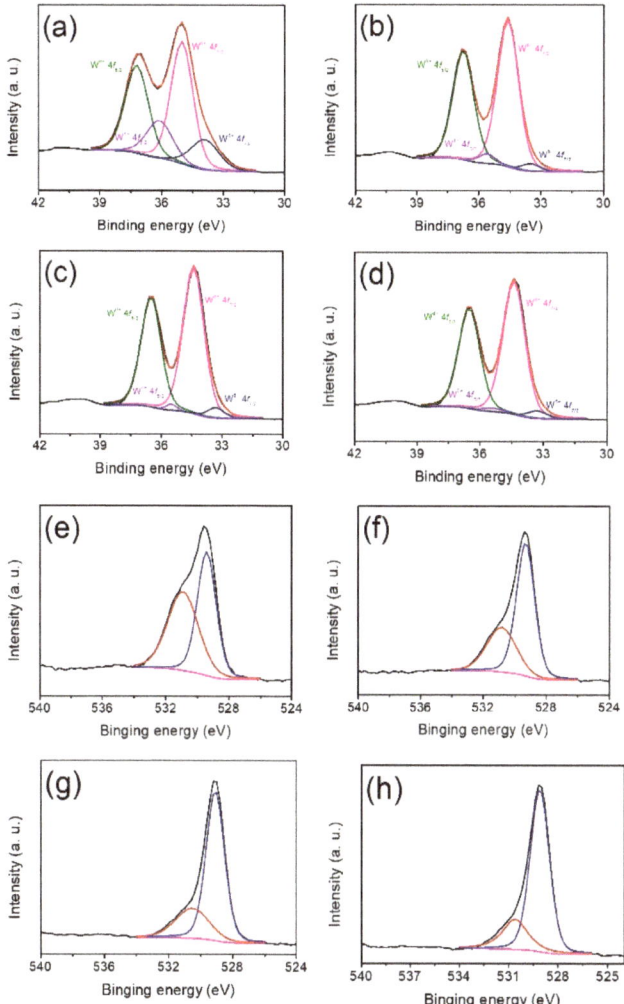

Figure 6. W4f XPS narrow scan spectra of the WO$_3$ thin films with and without annealing: (**a**) as-grown, (**b**) 400 °C, (**c**) 500 °C, (**d**) 600 °C. O1s XPS narrow scan spectra of the WO$_3$ thin films with and without annealing: (**e**) as-grown, (**f**) 400 °C, (**g**) 500 °C, (**h**) 600 °C.

The transmittance spectra of the WO$_3$ thin films with and without annealing are demonstrated in Figure 7a. The light was highly absorbed in the visible region with less than 40% transmittance for the as-grown WO$_3$ film, attributed to the presence of massive oxygen-related defects. Moreover, the as-grown WO$_3$ film is seen in semi-transparent bluish color, which shows the amorphous and highly non-stoichiometric natures of the film [18]. The highly transparent feature was observed for the WO$_3$ films conducted with annealing; moreover, no blue colouration appeared in the samples. The enhancement in the transmittance degree of the annealed WO$_3$ films is due to the reduction of oxygen-related crystal defects, which might play an important role in scattering the incident light. Notably, a clear shoulder feature appeared at approximately 350 nm for the films annealed below 600 °C. That shoulder feature in the optical transmittance spectra is associated with the residual crystal defects associated with oxygen deficiencies in the samples [19]. Notably, the shoulder feature completely vanished for the WO$_3$ film conducted with thermal annealing at 600 °C, indicating that the oxygen-related crystal defects of the film were substantially removed in the annealing process. The bandgap value of various thin-film samples is calculated by plotting $(\alpha h\nu)^{1/2}$ vs. photon energy using the following formula:

$$\alpha h\nu = A(h\nu - E_g)^n \tag{1}$$

where α is the absorption coefficient, A is a constant, $h\nu$ is the energy of an incident photon, E_g is the bandgap value. On extrapolating the linear portion of the curves (Figure 7b), the intercept on the energy axis $(\alpha h\nu)^{1/2} = 0$ gives the value of the indirect bandgap energy. The bandgap values were calculated as 2.48, 3.04, 3.04, and 2.78 eV for as-grown and annealed WO$_3$ samples at 400, 500, and 600 °C, respectively. The as-prepared WO$_3$ thin film exhibited the smallest bandgap value and may be associated with a relatively large content of oxygen deficiencies in the film as compared to the annealed ones. Due to the existence of a high density of oxygen-deficient crystal defects, they might form new discrete energy bands below the conduction band, resulting in the relatively low band gap. After annealing at 400 and 500 °C, the E_g of two WO$_3$ thin-film samples originates from the recombination of free carriers from the bottom of the conduction band energy to the valence band energy with the decreased discrete energy bands after thermal annealing. It was noticed that the bandgap value of the WO$_3$ film further decreased at the highest annealing temperature of 600 °C, assigned to explosive growth in grain size of the WO$_3$ film [20].

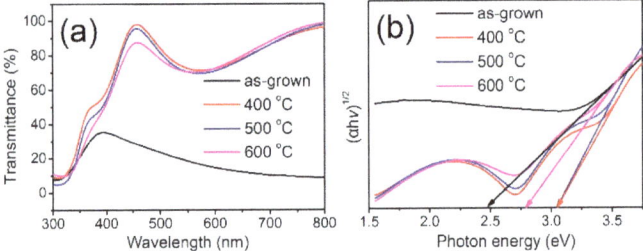

Figure 7. (**a**) Optical transmittance spectra of various WO$_3$ films, (**b**) Tauc plot of various WO$_3$ films.

Figure 8a–d show the absorbance spectra of the MB solution in the presence of various WO$_3$ thin films at different irradiation times. After the MB solution was illuminated, the absorbance peak intensity at ~663 nm was observed to gradually decrease with duration, implying that MB molecules are photodegraded. The photodegradation degrees (C/C_0) of the MB solution containing various WO$_3$ thin films are summarized in Figure 7e. The C_0 is concentration MB solution without irradiation and C is the residual concentration of the MB solution after irradiation at a given duration. Notably, the C/C_0 values of the MB solution containing various WO$_3$ thin films under various dark conditions are demonstrated in Figure 7e. A slightly decrease of C/C_0 value was observed for the dark balance of 120 min; this is attributed to the fact that partial MB molecules were absorbed on the surface of the WO$_3$ thin films under dark balance condition. By contrast, the photodegradation rates of the MB solution with various WO$_3$ thin films are different. As demonstrated in Figure 8e, the WO$_3$ film conducted with thermal annealing at 600 °C was most catalytically efficient, giving a photodegradation extent of approximately 45% in 30 min, while other annealed thin films photodegraded only 35% of MB in the same time period. After 120 min irradiation, the WO$_3$ film annealed at 600 °C still displayed the largest degree of photodegradation toward the MB solution. Notably, under irradiation, WO$_3$ was photoexcited and the e^-/h^+ pairs were formed. The e^- can participate in organic pollutant degradation reactions. The possible formations of the superoxide anion, hydroperoxyl, and hydrogen peroxide (H$_2$O$_2$) species in the organic dye solution are advantageous for further degrading the MB dyes [21]. However, it has been shown that the position of the conduction band of WO$_3$ (+0.50 V vs. NHE) was below the standard redox potential for the formation of superoxide anion (-0.33 V vs. NHE) and hydroperoxyl (−0.046 V vs. NHE) [22]. Based on the aforementioned, the following reactions are therefore more likely to occur during the MB photodegradation process using WO_3 thin films as photocatalysts:

$$H_2O_2 + e^- \rightarrow \bullet OH + OH^- \qquad (2)$$

$$H_2O_2 + h^+ \rightarrow \bullet OOH + H^+ \qquad (3)$$

$$H_2O_2 + \bullet OOH \rightarrow \bullet OH + H_2O + O_2 \qquad (4)$$

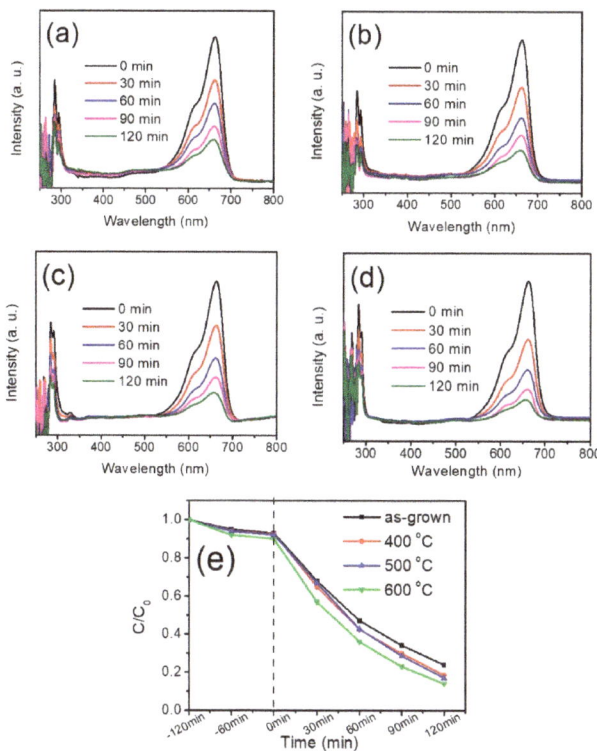

Figure 8. The intensity variation of absorbance spectra of the MB solution in presence of various WO_3 thin films under different irradiation durations: (**a**) as-grown, (**b**) 400 °C, (**c**) 500 °C, (**d**) 600 °C, (**e**) Plot of C/C_0 vs. irradiation time.

Meanwhile, the photogenerated h^+ in the WO_3 might involve the reactions and form $\bullet OH$ radicals through the following equations:

$$H_2O + h^+ \rightarrow H^+ + \bullet OH \qquad (5)$$

$$OH^- + h^+ \rightarrow \bullet OH \qquad (6)$$

The produced •OH radicals are efficiently degrading species for the MB dyes. Although the band position of the WO_3 is advantageous for the photodegradation mechanism, the microstructure and optical properties should be considered for the final photodegradation efficiency. The WO_3 film with thermal annealing performed at 600 °C was the most active photocatalyst, and the films annealed at lower temperatures were somewhat less active in this work. In contrast, the as-grown WO_3 film had a lower photocatalytic activity. We assume that the deteriorated crystal quality of the WO_3 with the lower annealing temperatures or without annealing was the decisive factor in their inferior photodegradation activity. Fewer oxygen-deficient-related defects in the lattice of the WO$_3$ film annealed at 600 °C might result in fewer recombination centers in the film, which would be detrimental for the higher photodegradation efficiency in this study. Moreover, the relative lower optical bandgap value of the WO$_3$ film conducted with thermal annealing at 600 °C among different annealed thin films is another advantageous factor to increase the light harvesting and enhance the degradation ability of crystalline WO$_3$ film toward MB dyes under irradiation.

4. Conclusions

The WO$_3$ thin films were DC sputtering deposited at 375 °C; moreover, the as-grown films exhibited an amorphous structure because of large composition deviation from the stoichiometric value. The as-grown WO$_3$ thin films were further conducted by thermal annealing procedures at 400–600 °C in ambient air. Structural analyses revealed that the amorphous WO$_3$ thin films crystallized after thermal annealing and demonstrated an orthorhombic structure. The surface grain size and surface roughness of the WO$_3$ films increased with annealing temperature. Moreover, the density of oxygen-deficiency-related crystal defects in the WO$_3$ films decreased with annealing temperature. The optical bandgap of WO$_3$ thin films are highly associated with the crystal quality and this can be controlled by conducting with different thermal annealing procedures. The as-grown WO$_3$ thin film annealed at 600 °C exhibited the highest photodegradation ability toward organic dyes in this study because of its high crystallinity, low crystal defects, and low optical bandgap among the various WO$_3$ thin films herein.

Author Contributions: Formal Analysis, C.-W.C.; Investigation, Y.-C.L. and C.-W.C.; Writing—Original Draft Preparation, Y.-C.L. and C.-W.C.; Writing—Review and Editing, Y.-C.L.; Supervision, Y.-C.L.

Funding: This research was funded by the Ministry of Science and Technology of Taiwan (MOST 105-2628-E-019-001-MY3).

Conflicts of Interest: The authors declare no conflict of interest.

References

1. Dong, P.; Hou, G.; Xi, X.; Shao, R.; Dong, F. WO$_3$-based photocatalysts: Morphology control, activity enhancement and multifunctional applications. *Environ. Sci. Nano* **2017**, *4*, 539–557. [CrossRef]
2. Lee, W.H.; Lai, C.W.; Hamid, S.B.A. One-step formation of WO$_3$-loaded TiO$_2$ nanotubes composite film for high photocatalytic performance. *Materials* **2015**, *8*, 2139–2153. [CrossRef]
3. Faudoa-Arzate, A.; Arteaga-Durán, A.; Saenz-Hernández, R.J.; Botello-Zubiate, M.E.; Realyvazquez-Guevara, P.R.; Matutes-Aquino, J.A. HRTEM microstructural characterization of β-WO$_3$ thin films deposited by reactive RF magnetron sputtering. *Materials* **2017**, *10*, 200. [CrossRef] [PubMed]
4. Shinguu, H.; Bhuiyan, M.M.H.; Ikegami, T.; Ebihara, K. Preparation of TiO$_2$/WO$_3$ multilayer thin film by PLD method and its catalytic response to visible light. *Thin Solid Films* **2006**, *506*, 111–114. [CrossRef]
5. Hunge, Y.M.; Mahadik, M.A.; Kumbhar, S.S.; Mohite, V.S.; Rajpure, K.Y.; Deshpande, N.G.; Moholkar, A.V.; Bhosale, C.H. Visible light catalysis of methyl orange using nanostructured WO$_3$ thin films. *Ceram. Int.* **2016**, *42*, 789–798. [CrossRef]
6. Arfaoui, A.; Touihri, S.; Mhamdi, A.; Labidi, A.; Manoubi, T. Structural, morphological, gas sensing and photocatalytic characterization of MoO$_3$ and WO$_3$ thin films prepared by the thermal vacuum evaporation technique. *Appl. Surf. Sci.* **2015**, *357*, 1089–1096. [CrossRef]
7. Dobromir, M.; Apetrei, R.P.; Rebegea, S.; Manole, A.V.; Nica, V.; Luca, D. Synthesis and characterization of RF sputtered WO$_3$/TiO$_2$ bilayers. *Surf. Coat. Technol.* **2016**, *285*, 197–202. [CrossRef]
8. Subrahmanyam, A.; Karuppasamy, A. Electrochromism. Optical and electrochromic properties of oxygen sputtered tungsten oxide (WO$_3$) thin films. *Sol. Energy Mater. Sol. Cells* **2007**, *91*, 266–274. [CrossRef]
9. Kim, T.S.; Kim, Y.B.; Yoo, K.S.; Sung, G.S.; Jung, H.J. Sensing characteristics of dc reactive sputtered WO$_3$ thin films as an NO$_x$ gas sensor. *Sens. Actuators B Chem.* **2000**, *62*, 102–108. [CrossRef]

10. Johansson, M.; Niklasson, G.; Österlund, L. Structural and optical properties of visible active photocatalytic WO$_3$ thin films prepared by reactive dc magnetron sputtering. *J. Mater. Res.* **2012**, *27*, 3130–3140. [CrossRef]
11. Stolze, M.; Gogova, D.; Thomas, L.-K. Analogy for the maximum obtainable colouration between electrochromic, gasochromic, and electrocolouration in DC-sputtered thin WO$_{3-y}$ films. *Thin Solid Films* **2005**, *476*, 185–189. [CrossRef]
12. Warren, B.E. *X-ray Diffraction*; Addison Wesley Publishing Co.: London, UK, 1969.
13. Liu, H.; Lu, H.; Zhang, L.; Wang, Z. Orientation selection in MgO thin films prepared by ion-beam-deposition without oxygen gas present. *Nucl. Instrum. Methods Phys. Res. B* **2015**, *360*, 60–63. [CrossRef]
14. Masudy-Panah, S.; Moakhar, R.S.; Chua, C.S.; Kushwaha, A.; Wong, T.I.; Dalapati, G.K. Rapid thermal annealing assisted stability and efficiency enhancement in a sputter deposited CuO photocathode. *RSC Adv.* **2016**, *6*, 29383–29390. [CrossRef]
15. Liang, Y.C.; Liang, Y.C. Physical properties of low temperature sputtering-deposited zirconium-doped indium oxide films at various oxygen partial pressures. *Appl. Phys. A* **2009**, *97*, 249–255. [CrossRef]
16. Hussain, T.; Al-Kuhaili, M.F.; Durrani, S.M.A.; Qurashi, A.; Qayyum, H.A. Enhancement in the solar light harvesting ability of tungsten oxide thin films by annealing in vacuum and hydrogen. *Int. J. Hydrog. Energy* **2017**, *42*, 28755–28765. [CrossRef]
17. Nayak, A.K.; Ghosh, R.; Santra, S.; Guha, P.K.; Pradhan, D. Hierarchical nanostructured WO$_3$–SnO$_2$ for selective sensing of volatile organic compounds. *Nanoscale* **2015**, *7*, 12460–12473. [CrossRef] [PubMed]
18. Castro-Hurtado, I.; Tavera, T.; Yurrita, P.; Pérez, N.; Rodriguez, A.; Mandayo, G.G.; Castaño, E. Structural and optical properties of WO$_3$ sputtered thin films nanostructured by laser interference lithography. *Appl. Surf. Sci.* **2013**, *276*, 229–235. [CrossRef]
19. Bujji Babu, M.; Madhuri, K.V. Structural, morphological and optical properties of electron beam evaporated WO$_3$ thin films. *J. Taibah Univ. Sci.* **2017**, *11*, 1232–1237.
20. Jain, P.; Arun, P. Influence of grain size on the band-gap of annealed SnS thin films. *Thin Solid Films* **2013**, *548*, 241–246. [CrossRef]
21. Aslam, I.; Cao, C.; Khan, W.S.; Tanveer, M.; Abid, M.; Idrees, F.; Riasat, R.; Tahir, M.; Butt, F.K.; Ali, Z. Synthesis of three-dimensional WO$_3$ octahedra: Characterization, optical and efficient photocatalytic properties. *RSC Adv.* **2014**, *4*, 37914–37920. [CrossRef]
22. Ghosh, A.; Mitra, M.; Banerjee, D.; Mondal, A. Facile electrochemical deposition of Cu$_7$Te$_4$ thin films with visible-light driven photocatalytic activity and thermoelectric performance. *RSC Adv.* **2016**, *6*, 22803–22811. [CrossRef]

© 2019 by the authors. Licensee MDPI, Basel, Switzerland. This article is an open access article distributed under the terms and conditions of the Creative Commons Attribution (CC BY) license (http://creativecommons.org/licenses/by/4.0/).

Article

Epitaxial Versus Polycrystalline Shape Memory Cu-Al-Ni Thin Films

Doga Bilican [1], Samer Kurdi [2], Yi Zhu [2], Pau Solsona [1], Eva Pellicer [1], Zoe H. Barber [2], Alan Lindsay Greer [2], Jordi Sort [1,3] and Jordina Fornell [1,*]

[1] Departament de Física, Facultat de Ciències, Universitat Autònoma de Barcelona, E-08193 Bellaterra (Cerdanyola del Vallès), Spain; bilican.doga@gmail.com (D.B.); pau.solsona@uab.cat (P.S.); Eva.pellicer@uab.cat (E.P.); jordi.sort@uab.cat (J.S.)
[2] Department of Materials Science and Metallurgy, University of Cambridge, Cambridge CB3 0FS, UK; sk862@cam.ac.uk (S.K.); yz475@cam.ac.uk (Y.Z.); zb10@cam.ac.uk (Z.H.B.); alg13@cam.ac.uk (A.L.G.)
[3] Institució Catalana de Recerca i Estudis Avançats (ICREA), Pg. Lluís Companys 23, E-08010 Barcelona, Spain
* Correspondence: jordina.fornell@uab.cat; Tel.: +34-9358-114-01

Received: 9 April 2019; Accepted: 6 May 2019; Published: 8 May 2019

Abstract: In this work, two different approaches were followed to obtain Cu-Al-Ni thin films with shape memory potential. On the one hand, Cu-Ni/Al multilayers were grown by magnetron sputtering at room temperature. To promote diffusion and martensitic/austenitic phase transformation, the multilayers were subjected to subsequent heat treatment at 800 °C and quenched in iced water. On the other hand, Cu, Al, and Ni were co-sputtered onto heated MgO (001) substrates held at 700 °C. Energy-dispersive X-ray spectroscopy, X-ray diffraction, and transmission electron microscopy analyses were carried out to study the resulting microstructures. In the former method, with the aim of tuning the thin film's composition, and, consequently, the martensitic transformation temperature, the sputtering time and applied power were adjusted. Accordingly, martensitic Cu-14Al-4Ni (wt.%) and Cu-13Al-5Ni (wt.%) thin films and austenitic Cu-12Al-7Ni (wt.%) thin films were obtained. In the latter, in situ heating during film growth led to austenitic Cu-12Al-7Ni (wt.%) thin films with a (200) textured growth as a result of the epitaxial relationship MgO(001)[100]/Cu-Al-Ni(001)[110]. Resistance versus temperature measurements were carried out to investigate the shape memory behavior of the austenitic Cu-12Al-7Ni (wt.%) thin films produced from the two approaches. While no signs of martensitic transformation were detected in the quenched multilayered thin films, a trend that might be indicative of thermal hysteresis was encountered for the epitaxially grown thin films. In the present work, the differences in the crystallographic structure and the shape memory behavior of the Cu-Al-Ni thin films obtained by the two different preparation approaches are discussed.

Keywords: Cu-Al-Ni; shape memory alloys; thin film; sputtering; size effects

1. Introduction

Shape memory alloys (SMA) exhibit displacive and reversible deformation behavior due to sensing thermodynamic and mechanical changes in their environment [1–7]. This plays a role in the development of components that can be cycled between two macroscopic shapes depending on temperature change. The fact that SMAs exhibit temperature-induced strain recovery makes them a type of advanced engineering material. Typical application fields for these materials are encountered in sensing–actuating systems in automotive, aerospace, robotics, and biomedical technologies [8]. Of the most functional shape memory alloys, Ni-Ti [9–11] Cu-Zn-Al [12], and Cu-Al-Ni [13,14] are some of the most widely used. Even though the Ni-Ti SMA system is widely studied and commercialized on account of its high percentage of shape recovery, Cu-based SMAs have become long-term proposed substitutions for the Ni-Ti system due to the fact that they have lower production costs in addition to having desirable properties such as a large superelastic effect, wide transformation temperature ranges, small hysteresis, and a high damping coefficient. Despite the fact that Fe-based SMAs, such as Fe-Mn and Fe-Mn-Si, also appear as substitute candidates for Ni-Ti due to their good workability and cost efficiency advantages, Fe-based SMAs undergo a large transformation hysteresis which limits their area for shape memory applications compared to Cu-based SMAs [15].

Regardless of the alloy composition, because of possessing large amounts of thermal capacitance, the application of shape memory behavior in bulk alloys is challenging. However, due to high actuation outputs per unit volume, the response time can be reduced substantially and the speed of operation may be increased sufficiently in shape memory thin films. As a consequence, these materials can be utilized in micro/nanorobotic platforms [16].

Rather than conventional elastic or plastic dislocation glide, the behavior of shape memory depends on displacive, diffusionless phase transformation, which takes place between a high temperature phase (austenite) and a low temperature one (martensite). As a non-equilibrium phase, the most commonly applied way to induce martensite in materials is to subject them to quenching after holding them at high temperature in order for the atoms to become locked into position before they reach their equilibrium states, with the lowest free energy [17]. Post-quenching is also convenient for introducing the shape memory effect in films grown at room temperature. Investigations on the synthesis of Fe-Pd [18], Ni-Ti [19], and Cu-Al-Ni [20] by sputter depositing and subsequent quenching have been reported in the literature. In thin film studies, epitaxy may be induced via growth on a single-crystal substrate, which dictates the crystal growth direction of the film. As a refractory, electrically insulating, and transparent substrate, MgO offers a flat interface for epitaxial growth. It shows a convenient lattice that matches many metals with face-centered cubic symmetry [21].

For thin film synthesis using in situ heating, epitaxial growth can be provided by techniques such as laser ablation [22], molecular beam epitaxy [23,24], and electron beam evaporation [25]. Since preferential orientation is favored, the enhancement of shape memory properties can be achieved. Among the sputtered shape memory alloys, Ni-Ti [26] and Ni_2MnGa [27,28] are the systems for which in-depth research on epitaxial growth is being carried out.

Cu-Al-Ni SMAs are known for their good thermal and electrical conductivity, broad interval of transformation temperatures (between 70 and 470 K), good thermal stability of martensitic transformation, and large recoverable strains [29–31]. They typically have Ni and Al contents between 2–5 wt.% and 13–15 wt.%, respectively [32,33]. Cu-Al-Ni SMAs have some advantages over the widely used Ni-Ti SMAs, such as a lower melting temperature, which facilitates composition control [34]. The fact that they have better corrosion resistance and a lower cost compared to Ni-Ti, which is prone to oxidation, makes Cu-Al-Ni components favorable choices [35] for certain applications. Furthermore, it has been shown that achieving further enhancement in the microstructure and mechanical properties of Cu-Al-Ni SMAs is possible via the minor addition of cobalt to its composition [34]. The structural and mechanical properties of the Cu-Al-Ni system have been studied both for bulk alloys, which are produced by methods such as rapid solidification, casting [36], and powder metallurgy [37], and thin films, which are produced by methods including electron beam evaporation [38], thermal evaporation [39], and sputtering [40,41].

Here, the dependence of the crystallographic phase structure on the chemical composition of the Cu-Al-Ni shape memory system is reported for the first time for sputtered thin films. Different compositions were synthesized by adjusting the Al and Ni content. Additionally, the crystallographic differences due to the undertaken synthetic approach were investigated. Two methods were implemented to produce the thin films. Firstly, free-standing films with varying compositions were prepared by sputtering multilayers of Cu-Ni and Al. The as-grown films were conventionally annealed at 800 °C and quenched in iced water, as would be done in a top-down approach, in order to obtain a set of samples with different chemical compositions: Cu-14Al-4Ni (wt.%), Cu-13Al-5Ni (wt.%) and Cu-12Al-7Ni (wt.%). Secondly, Cu-12Al-7Ni (wt.%) thin films were grown on MgO(001) substrates held at 700 °C, taking advantage of the epitaxial relationship between MgO(001) and austenitic β-Cu-Al-Ni as a bottom-up approach. Unlike most of the previous studies for sputtered Cu-Al-Ni samples, where a Cu-Al-Ni target was used for synthesis [16,27,34], in this work all depositions used individual Cu, Al, and Ni targets (i.e., co-sputtering). Structural characterization studies were carried out to investigate the crystallographic properties of the films. The martensitic phase transformation behavior of the samples was also investigated.

2. Materials and Methods

Cu, Al and Ni targets of 99.95% purity were used to grow all samples. In order to produce free-standing Cu-Al-Ni films, Si substrates were covered with resin by spin coating before sputtering. The deposition of multilayers was carried out by DC sputtering from Cu and Al targets and RF sputtering from a Ni target. Seven multilayers consisting of three Al films deposited between Cu-Ni films were prepared (Figure 1a). After the deposition, the samples were rinsed first in acetone and later in ethanol to dissolve the resin and to separate the film from the substrate. In order to promote diffusion to obtain a homogeneous composition, the films, wrapped in Ta foils, were subjected to annealing at 800 °C for 60 min in a sealed quartz tube and subsequently quenched in iced water.

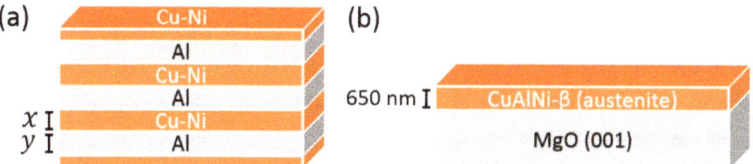

Figure 1. Sketch of the (**a**) prepared Cu-Ni/Al multilayers before quenching (for x and y values refer to Table 1) and (**b**) Cu-Al-Ni film on (001) oriented MgO.

Table 1. Thickness values of the sputtered multilayers before quenching. The top and bottom Cu–Ni layers have half the thickness of the layers in between.

Composition	Layer Thickness (nm)		Total Thin Film Thickness (nm)
	Cu-Ni (x)	Al (y)	Cu-Al-Ni
Cu-14Al-4Ni (wt.%)	230	58	864
Cu-13Al-5Ni (wt.%)	242	63	915
Cu-12Al-7Ni (wt.%)	347	70	1250

The top and bottom Cu-Ni layers had half the thickness of the Cu-Ni layers in between (Figure 1a). The details of the thickness values for each composition are given in Table 1, while sputtering conditions are listed in Table 2.

Table 2. The power (W) and time (min) conditions applied for the sputtering of samples Cu–14Al–4Ni (wt.%), Cu–13Al–5Ni (wt.%), and Cu–12Al–7Ni (wt.%).

Composition	Cu-Ni			Al	
	P (W)		t (min)	P (W)	t (min)
	Cu	Ni			
Cu-14Al-4Ni (wt.%)	200	60	10	200	14
Cu-13Al-5Ni (wt.%)	200	75	9	200	15.5
Cu-12Al-7Ni (wt.%)	200	85	10	200	17

The co-sputtered 650 nm thick Cu-12Al-7Ni (wt.%) thin film was deposited onto a 700 °C heated MgO(001) substrate (Figure 1b) from individual Cu, Al, and Ni targets using powers of 22, 30, and 5 W, respectively. The deposition pressure was set to 0.65 Ar, and the deposition rate was approximately 4.3 nm/min.

Compositional analyses were performed with a field emission scanning electron microscope (FE-SEM, Zeiss, Oberkochen, Germany) equipped with an energy dispersive X-ray spectroscopy (EDX) detector operated at 15 kV. To confirm the homogeneity of the film along its thickness, EDX analysis was also carried out on the film cross-section. Structural characterization was carried out by X-ray diffraction (XRD, PANalytical, Royston, UK) (θ/2θ diffraction with Cu Kα radiation) and transmission electron microscopy (TEM, JEOL 2011 200 KV, Peabody, MA, USA). For TEM observations, the free-standing films were prepared with a GATAN polishing ion device (Pleasanton, CA, USA) while the Cu-12Al-7Ni (wt.%) film grown on MgO was scratched from the substrate and milled with an agate mortar. A Brucker D8 theta/theta four circle diffractometer (Billerica, MA, USA): Ω, 2θ, X, Φ, graded mirror (GM; to give a nearly parallel beam) equipped with a scintillation counter detector was used to study the epitaxial films grown on MgO. Coating thickness measurements were made using the 3D optical surface metrology system, Leica DCM 3D (Leica Microsystems Inc., Buffalo Grove, IL, USA). In order to investigate martensitic transformation behavior, electrical resistance measurements were performed with a 2 K/min heating rate.

3. Results and Discussion

The XRD patterns of films synthesized by both routes are shown in Figure 2a. Among the free-standing thin films produced by multilayer deposition, Cu-14Al-4Ni (wt.%) (Figure 2a) showed a mixture of β' martensite and pure Cu. β' martensite had a monoclinic structure (space group Cmcm), whereas Cu had a cubic Fm-3m lattice. As the content of Ni increased at the expense of Al, traces of β austenite (space group Fm-3m) along with β' martensite were detected for Cu-13Al-5Ni (wt.%) (Figure 2b), and only β austenite was seen in the XRD pattern of Cu-12Al-7Ni (wt.%) (Figure 2c). In turn, the XRD pattern of the thin film co-sputtered at a high temperature on MgO mainly consisted of textured β austenite, but the Cu_9Al_4 phase (space group P-43m) was also present (Figure 2d). Previous studies on Cu-Al-Ni bulk shape memory alloys have reported that a slight change in composition results in a shift in the transformation temperatures [33,41,42]. The studies carried out by Recarté et al. revealed that, for a fixed aluminum content of 13.2 wt.%, a decrease in nickel from 5 wt.% to 3.5 wt.% raised the austenite finish temperature from 10 to 80 °C [33]. Agafonov et al. showed that Al content variation from 14.98 wt.% to 13.03 wt.% caused a change in the room temperature phase from austenite to martensite for samples synthesized by casting and quenching in water [41]. Similar trends were observed by Suresh and Ramamurty [42] where, at room temperature, a Cu-13.4Al-4Ni (wt.%) alloy was in the martensite state, but a Cu-14.1Al-4Ni (wt.%) alloy was austenitic. In our work, we observed that a change in Ni content followed the same trend in the transformation temperatures as the results reported by Recarté et al. [33], but the opposite tendency than that reported in [33,41] was observed when the aluminum content was modified. This may be because in our case, we were modifying Ni and Al content simultaneously, and, consequently, our transformation temperatures were influenced by both Al and Ni.

The film grown on MgO(001) showed a Cu-12Al-7Ni (wt.%) composition. The epitaxial relationship MgO(001)[100]/Cu-Al-Ni(001)[110] was induced. The film grew with a 45° in-plane rotation on the cubic cell of the substrate. The lattice mismatch (f) between the cubic lattices is shown in Equation (1):

$$f = (\sqrt{2}a_{MgO} - a_{CuAlNi\ \beta})/a_{CuAlNi\ \beta} \tag{1}$$

where a_{MgO} = 4.212 Å, $a_{CuAlNi\ \beta}$ = 5.836 Å.

The lattice mismatch between the film and the substrate was f = 2%, which was small enough to favor epitaxy. The epitaxial film, grown at 700 °C, exhibited preferential (100) out-of-plane orientation. This showed up as strong (200) and (400) peaks in the XRD pattern (Figure 2d). Peaks belonging to the Cu_9Al_4 phase (i.e., (200), (421), (332), and (550) planes) were also present. The formation of α-Cu and Cu_9Al_4 phases is typically observed in Cu-Al-Ni systems [43,44]. The β phase in the Cu-Al-Ni system undergoes a eutectoid decomposition at ~840 K into α (Cu) and γ2 (Cu_9Al_4), then, the stable phases at room temperature are Cu and Cu$_9$Al$_4$. However, if the alloy is quenched at a sufficiently high cooling rate from the β phase region to ambient temperature, the β phase may be retained or it may transform martensitically. However, if the cooling rate is not high enough, traces of Cu or Cu$_9$Al$_4$ may be present in the alloy.

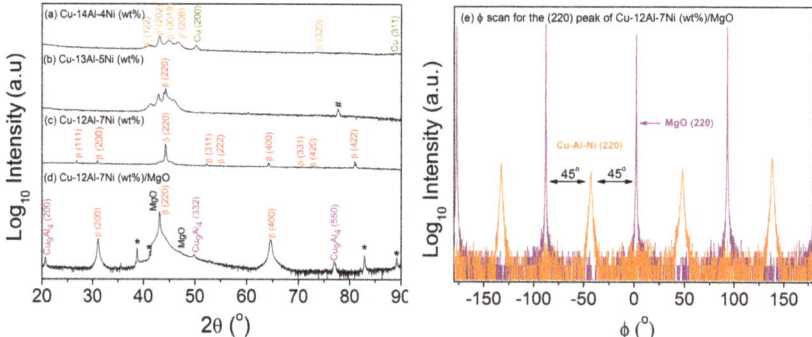

Figure 2. XRD patterns of the sputtered Cu–Al–Ni thin films: (**a**) θ/2θ scan for Cu–14Al–4Ni (wt.%); (**b**) θ/2θ scan for Cu–13Al–5Ni (wt.%); (**c**) θ/2θ Cu–12Al–7Ni (wt.%) films obtained by multilayer sputter deposition and subsequent heat treatment; (**d**) θ/2θ scan for Cu–12Al–7Ni (wt.%) film prepared on MgO substrate at 700 °C; and (**e**) ϕ scan for the (220) peak of the β–Cu–Al–Ni film deposited at 700 °C on MgO(001). # peak is unidentified. * peaks originate from the sample holder.

The phi scan carried out for the (220) peak of this film confirmed that the film lattice was rotated 45° in-plane, relative to the substrate (Figure 2e).

For the free-standing thin films produced by multilayer deposition, further evidence that the compositional change from Cu-14Al-4Ni (wt.%) to Cu-12Al-7Ni (wt.%) resulted in a shift from martensite to austenite was observed by TEM (Figure 3). Figure 3a,c shows TEM images of Cu-14Al-4Ni (wt.%) and Cu-12Al-7Ni (wt.%) thin films, respectively. Typical martensitic plates were present in Cu-14Al-4Ni, while a regular polycrystalline structure was observed in the Cu-12Al-7Ni (wt.%) alloy. The Selected Area Electron Diffraction (SAED) pattern from Cu-14Al-4Ni (wt.%) thin film (Figure 3b) consisted of diffraction spots that belong to β′ martensite corresponding to the (040), (202), (0018), (202), and (122) planes and to Face-Centered Cubic (FCC) Cu (i.e., (220), (311), (311), and (033) planes). Conversely, the SAED pattern of the Cu-12Al-7Ni (wt.%) thin film (Figure 3d) consisted of diffraction spots characteristic of the austenitic phase (β) corresponding to (111), (511), (711), (622), (422), and (533) planes. In Figure 4a, a high resolution TEM image of the epitaxial Cu-12Al-7Ni (wt.%) thin film grown on MgO is shown. From the fast-Fourier transform (FFT) pattern of the selected zone shown in Figure 4a, an interplanar distance value of 2.07 Å (which belongs to β austenite (220)), as well as a distance of 2.93 Å (which belongs to β austenite (200)), was identified (Figure 4b).

In order to identify the transformation temperatures, electrical resistance measurements were carried out as a function of temperature for both the quenched multilayered sample with the composition Cu-12Al-7Ni (wt.%) and the sample with the same composition grown on MgO. Martensitic transformation was not observed in the quenched multilayered Cu-12Al-7Ni (wt.%) (Figure 5a), whereas a trend that might be indicative of transformation hysteresis was found for the epitaxially grown Cu-12Al-7Ni (wt.%) sample (Figure 5b). The reason behind the fact that a transformation was observed in the epitaxially grown (200) textured austenite thin film whereas no change was seen in the randomly oriented polycrystalline one could be related to size effects. Size effects, such as the volume of the material or structural components including precipitate particles and grains in polycrystals, have a huge influence on the martensitic phase transformation [37]. Decreasing grain size and decreasing twin separation cause an increase in strain energy and twin interfacial energy which, in turn, increases the energy barrier [38]. For instance, Shi et al. [39] showed that for submicrometric NiTi particles (below 100 nm), the martensitic transformation was fully suppressed. Similarly, in our work, no transformation was observed in the randomly oriented polycrystalline austenitic Cu-12Al-7Ni (wt.%) film with a crystallite size of 95 nm according to Scherrer's formula.

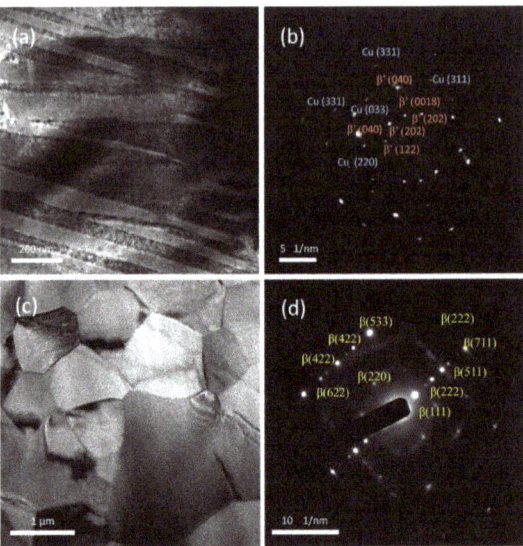

Figure 3. (a) Transmission electron microscopy (TEM) image of Cu–14Al–4Ni (wt.%) thin film produced from multilayer deposition followed by annealing; (b) SAED pattern of (a,c) TEM image of Cu–12Al–7Ni (wt.%) thin film; (d) SAED pattern of (c).

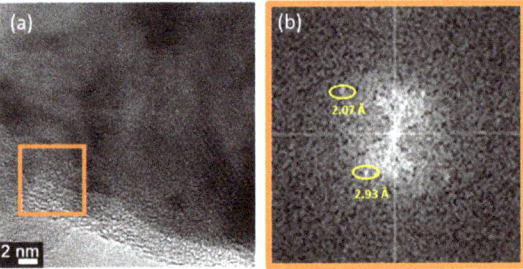

Figure 4. (a) HRTEM image of Cu–12Al–7Ni (wt.%) thin film grown on MgO; (b) FFT of the selected zone (orange square) in (a).

Additionally, previous studies demonstrate that decreasing film thickness is also a reason for the suppression of martensitic phase transformation [45–48]. Wan and Komvopoulos [48] showed that for sputtered NiTi films with a thickness of less than 100 nm, no martensitic transformation was observed. In the case of sputtered polycrystalline Cu-Al-Ni films, transformation temperatures for films with 2 μm thickness were reported by Moran and his co-workers [31]. To the best of our knowledge, these are the sputtered Cu-Al-Ni SMA films with the lowest thickness values reported in the literature. Torres et al. [20] also observed martensitic transformation for 5 μm thick films. In contrast, no transformation was observed in the polycrystalline 1.25 μm Cu-12Al-7Ni (wt.%) sample prepared in this work by post-treatment of the multilayers. It has been indicated by Chen and Schuh [49] that as d(grain size)/D(sample thickness) decreases, the energy barrier for transformation increases.

The interfacial energies depend on grain boundaries formed between martensite plates, the interfaces between austenite and martensite plates, and the twin interfaces within martensite plates [48].

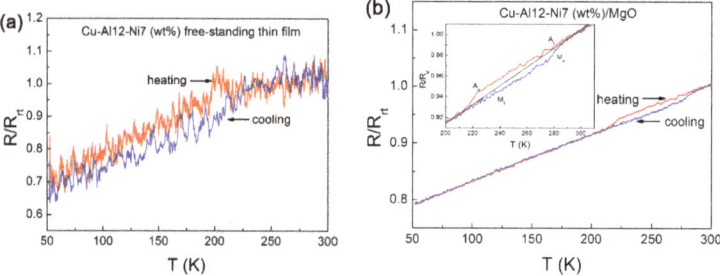

Figure 5. Resistance vs. temperature measurement of Cu–12Al–7Ni (wt.%) films prepared by (**a**) heat treatment after sputtering and (**b**) in situ heating during growth.

The absence of phase transformation in films formed at low deposition temperatures could be attributed to the small grain sizes causing a higher number of grain boundary interfaces and associated excess free volume, and/or the thinner nature of the samples, as noted by Wan and Komvopoulos [48].

4. Conclusions

In this work, structural characterization studies were carried out to investigate the crystallographic properties of Cu-Al-Ni films and their transformation temperatures.

The main conclusions from this work are as follows:

- The microstructure of Cu-Al-Ni sputtered films is found to depend both on the alloy composition as well as the experimental procedure used to grow the films.
- A transition from martensite to austenite was observed as the Ni content increased and the Al content decreased in samples prepared by multilayer sputtering followed by quenching.
- Preferential growth along the (100) direction was observed in β-austenite Cu-12Al-7Ni (wt.%) grown on MgO at 700 °C due to the epitaxial relationship MgO(001)[100]/Cu-Al-Ni(001)[110].
- Resistance change with respect to temperature, suggesting martensitic transformation hysteresis, was observed in the preferentially oriented austenitic Cu-12Al-7Ni (wt.%) film, whereas martensitic transformation was completely suppressed in the polycrystalline austenitic sample produced by multilayer sputtering with the same composition.

Author Contributions: Conceptualization, D.B. and J.F.; Methodology, D.B., P.S., S.K., Y.Z, and J.F.; Formal Analysis, D.B., S.K., and J.F.; Investigation, D.B., S.K., Z.H.B., and J.F.; Resources, D.B., S.K., Z.H.B., and J.F.; Data Curation, D.B. and J.F.; Writing—Original Draft Preparation, D.B.; Writing—Review and Editing, D.B., S.K., Z.H.B., A.L.G., E.P., J.S., and J.F.; Supervision, E.P., J.S., and J.F.; Project Administration, E.P., J.S., and J.F.; Funding Acquisition, E.P. and J.S.

Funding: This work was supported by the SELECTA (No. 642642) H2020-MSCA-ITN-2014 project. Partial financial support from the Spanish government (Project MAT2017-86357-C3-1-R and associated FEDER) and the Generalitat de Catalunya (2017-SGR-292) is acknowledged.

Acknowledgments: E.P. is grateful to MINECO for the "Ramon y Cajal" contract (RYC-2012-10839). J.F. acknowledges the "Juan de la Cierva" fellowship from MINECO (IJCI-2015-27030).

Conflicts of Interest: The authors declare no conflict of interest.

References

1. Otsuka, K.; Wayman, C.M. *Shape Memory Materials*; Cambridge University Press: Cambridge, UK, 1999.
2. Ma, J.; Karaman, I. Expanding the repertoire of shape memory alloys. *Science* **2010**, *327*, 1468–1469. [CrossRef]
3. Cingolani, E.; Ahlers, M.; Van Humbeeck, J. Stabilization and two-way shape memory effect in Cu-Al-Ni single crystals. *Met. Mater. Trans. A* **1999**, *30*, 493–499. [CrossRef]
4. Ishibashi, M.; Tabata, N.; Suetake, T.; Omori, T.; Sutou, Y.; Kainuma, R.; Yamauchi, K.; Ishida, K. A simple method to treat an ingrowing toenail with a shape-memory alloy device. *J. Dermatol.* **2008**, *19*, 291–292. [CrossRef] [PubMed]
5. Seelecke, S.; Muller, I. Shape memory alloy actuators in smart structures: Modeling and simulation. *Appl. Mech. Rev.* **2004**, *57*, 23–46. [CrossRef]
6. Jani, J.M.; Leary, M.; Subic, A.; Gibson, M.A. A review of shape memory alloy research, applications and opportunities. *Mater. Des.* **2014**, *56*, 1078–1113. [CrossRef]
7. Montero-Ocampo, C.; Lopez, H.; Salinas Rodriguez, A. Effect of compressive straining on corrosion resistance of a shape memory Ni-Ti alloy in ringer's solution. *J. Biomed. Mater. Res.* **1993**, *32*, 583–591. [CrossRef]
8. Bogue, R. Shape-memory materials: A review of technology and applications. *Assem. Autom.* **2009**, *29*, 214–219. [CrossRef]
9. Dehghanghadikolaei, A.; Ibrahim, H.; Amerinatanzi, A.; Hashemi, M.; Moghaddam, N.S.; Elahinia, M. Improving corrosion resistance of additively manufactured nickel-titanium biomedical devices. *J. Mater. Sci.* **2019**, *54*, 7333–7355. [CrossRef]
10. Eggeler, G.; Hornbogen, E.; Yawny, A.; Heckmann, A.; Wagner, M. Structural and functional fatigue of NiTi fhape memory alloys. *Mater. Sci. Eng. A* **2004**, *378*, 24–33. [CrossRef]
11. Ibrahim, H.; Jahadakbar, A.; Dehghan, A.; Moghaddam, N.S.; Amerinatanzi, A.; Elahinia, M. In vitro corrosion assessment of additively manufactured porous NiTi structures for bone fixation applications. *Metals* **2018**, *8*, 164. [CrossRef]
12. Perkins, J.; Sponholz, R.O. Stress-induced martensitic transformation cycling and two-way shape memory training in Cu-Zn-Al alloys. *Met. Mater. Trans. B* **1984**, *15*, 313–321. [CrossRef]
13. Ivanić, I.; Kožuh, S.; Grgurić, T.H.; Kosec, B.; Gojić, M. The influence of heat treatment on microstructure and phase transformation temperatures of Cu-Al-Ni shape memory alloys. *Kem. Ind.* **2019**, *68*, 111–118. [CrossRef]
14. Recarte, V.; Pérez-Landazábal, J.I.; Ibarra, A.; Nó, M.L.; San Juan, J. High temperature β phase decomposition process in a Cu-Al-Ni shape memory alloy. *Mater. Sci. Eng. A* **2004**, *378*, 238–242. [CrossRef]
15. Alaneme, K.K.; Okotete, E.A. Reconciling viability and cost-effective shape memory alloy options–A review of copper and iron based shape memory metallic systems. *Eng. Sci. Technol. Int. J.* **2016**, *19*, 1582–1592. [CrossRef]
16. Pan, Q.; Cho, C. The investigation of a shape memory alloy micro-damper for MEMS applications. *Sensors* **2007**, *7*, 1887–1900. [CrossRef]
17. Olson, G.B.; Cohen, M. A general mechanism of martensitic nucleation: Part I. General concepts and the FCC → HCP transformation. *Metall. Mater. Trans. A* **1976**, *7*, 1897–1904.
18. Sugimura, Y.; Cohen-Karni, I.; McCluskey, P.; Vlassak, J. Stress evolution in sputter-deposited Fe–Pd shape-memory thin films. *J. Mater. Res.* **2005**, *20*, 2279–2287. [CrossRef]
19. Chu, J.P.; Lai, Y.W.; Lin, T.N.; Wang, S.F. Deposition and characterization of TiNi-base thin films by sputtering. *Mater. Sci. Eng. A* **2000**, *277*, 11–17. [CrossRef]
20. Torres, C.E.; Condo, A.; Haberkorn, N.; Zelaya, E.; Schryvers, D.; Guimpel, J.; Lovey, F. Structures in textured Cu–Al–Ni shape memory thin films grown by sputtering. *Mater. Charact.* **2014**, *96*, 256–262. [CrossRef]
21. Tolstova, Y.; Omelchenko, S.T.; Shing, A.M.; Atwater, H.A. Heteroepitaxial growth of Pt and Au thin films on MgO single crystals by bias-assisted sputtering. *Sci. Rep.* **2016**, *6*, 23232. [CrossRef]
22. Gu, H.; You, L.; Leung, K.; Chung, C.; Chan, K.; Lai, J. Growth of TiNiHf shape memory alloy thin films by laser ablation of composite targets. *Appl. Surf. Sci.* **1998**, *127*, 579–583. [CrossRef]
23. Dong, J.W.; Chen, L.C.; Palmstro/m, C.J.; James, R.D.; McKernan, S. Molecular beam epitaxy growth of ferromagnetic single crystal (001) Ni_2MnGa on (001) GaAs. *Appl. Phys. Lett.* **1999**, *75*, 1443–1445. [CrossRef]

24. Shih, T.C.; Xie, J.Q.; Dong, J.W.; Dong, X.Y.; Srivastava, S.; Adelmann, C.; McKernan, S.; James, R.D.; Palmstrøm, C.J. Epitaxial growth and characterization of single crystal ferromagnetic shape memory Co$_2$NiGa films. *Ferroelectrics* **2006**, *342*, 35–42. [CrossRef]
25. Kühnemund, L.; Edler, T.; Kock, I.; Seibt, M.; Mayr, S.G. Epitaxial growth and stress relaxation of vapor-deposited Fe-Pd magnetic shape memory films electron beam evaporation. *New J. Phys.* **2009**, *11*, 113054. [CrossRef]
26. Gisser, K.R.C.; Busch, J.D.; Johnson, A.D.; Ellis, A.B. Oriented nickel-titanium shape memory alloy films prepared by annealing during deposition. *Appl. Phys. Lett.* **1992**, *61*, 1632–1634. [CrossRef]
27. Jenkins, C.A.; Ramesh, R.; Huth, M.; Eichhorn, T.; Pörsch, P.; Elmers, H.J.; Jakob, G. Growth and magnetic control of twinning structure in thin films of Heusler shape memory compound Ni$_2$MnGa. *Appl. Phys. Lett.* **2008**, *93*, 234101. [CrossRef]
28. Niemann, R.; Backen, A.; Kauffmann-Weiss, S.; Behler, C.; Rößler, U.; Seiner, H.; Heczko, O.; Nielsch, K.; Schultz, L.; Fähler, S. Nucleation and growth of hierarchical martensite in epitaxial shape memory films. *Acta Mater.* **2017**, *132*, 327–334. [CrossRef]
29. Goryczka, T. Effect of wheel velocity on texture formation and shape memory in Cu-Al-Ni melt-spun ribbons. *Arch. Metall. Mater.* **2009**, *54*, 755–763.
30. Delaey, L. Diffusionless transformations. In *Phase Tranformation in Materials*; Kostorz, G., Ed.; Wiley-VCH: Weinheim, Germany, 2001.
31. Morán, M.; Condó, A.; Soldera, F.; Sirena, M.; Haberkorn, N. Martensitic transformation in freestanding and supported Cu–Al–Ni thin films obtained at low deposition temperatures. *Mater. Lett.* **2016**, *184*, 177–180. [CrossRef]
32. Niedbalski, S.; Durán, A.; Walczak, M.; Ramos-Grez, J.A. Laser-assisted synthesis of Cu-Al-Ni shape memory alloys: Effect of inert gas pressure and Ni content. *Materials* **2019**, *12*, 794. [CrossRef]
33. Recarte, V.; Pérez-Sáez, R.B.; Juan, J.S.; Bocanegra, E.H.; Nó, M.L. Influence of Al and Ni concentration on the Martensitic transformation in Cu-Al-Ni shape-memory alloys. *Met. Mater. Trans. A* **2002**, *33*, 2581–2591. [CrossRef]
34. Braga, F.D.O.; Matlakhov, A.N.; Matlakhova, L.A.; Monteiro, S.N.; De Araújo, C.J. Martensitic transformation under compression of a plasma processed polycrystalline shape memory CuAlNi Alloy. *Mater. Res.* **2017**, *20*, 1579–1592. [CrossRef]
35. Haidar, M.A.; Saud, S.N.; Hamzah, E. Microstructure, mechanical properties, and shape memory effect of annealed Cu-Al-Ni-xCo shape memory alloys. *Metallogr. Microstruct. Anal.* **2018**, *7*, 57–64. [CrossRef]
36. Saud, S.N.; Bakar, T.A.A.; Hamzah, E.; Ibrahim, M.K.; Bahador, A. Effect of quarterly element addition of cobalt on phase transformation Cu-Al-Ni shape memory alloys. *Metall. Mater. Trans. A* **2015**, *46*, 3528–3542. [CrossRef]
37. Sharma, M.; Vajpai, S.K.; Dube, R.K. Processing and characterization of Cu-Al-Ni shape memory alloy strips prepared via a novel powder metallurgy route. *Metall. Mater. Trans. A* **2010**, *41*, 2905–2913. [CrossRef]
38. Gómez-Cortés, J.; Juan, J.S.; Lopez, G.A.; Nó, M. Synthesis and characterization of Cu–Al–Ni shape memory alloy multilayer thin films. *Thin Solid Films* **2013**, *544*, 588–592. [CrossRef]
39. Canbay, C.A.; Tekatas, A.; Ozkul, I. Fabrication of Cu-Al-Ni shape memory thin film by thermal evaporation. *Turk. J. Eng.* **2017**, *1*, 27–32.
40. Minemura, T.; Andoh, H.; Kita, Y.; Ikuta, I. Shape memory effect and microstructures of sputter-deposited Cu-Al-Ni films. *J. Mater. Sci. Lett.* **1985**, *4*, 793–796. [CrossRef]
41. Agafonov, V.; Naudot, P.; Dubertret, A.; Dubois, B. Influence of the aluminium content on the appearance and stability of martensites in the Cu Al Ni system. *Scr. Met.* **1988**, *22*, 489–494. [CrossRef]
42. Suresh, N.; Ramamurty, U. Aging response and its effect on the functional properties of Cu–Al–Ni shape memory alloys. *J. Alloy. Compd.* **2008**, *449*, 113–118. [CrossRef]
43. Sharma, M.; Vajpai, S.K.; Dube, R.K.; Vajpai, S. Synthesis and properties of Cu–Al–Ni shape memory alloy strips prepared via hot densification rolling of powder preforms. *Power Metall.* **2011**, *54*, 620–627. [CrossRef]
44. Araujo, A.P.M.; Simões, J.B.; Araújo, C.J. Analysis of compositional modification of commercial aluminum bronzes to obtain functional shape memory properties. *Mater. Res.* **2017**, *20*, 331–341. [CrossRef]
45. Malygin, G.A. Nanoscopic size effects on martensitic transformations in shape memory alloys. *Phys. Solid State* **2008**, *50*, 1538–1543. [CrossRef]
46. Waitz, T.; Karnthaler, H. Martensitic transformation of NiTi nanocrystals embedded in an amorphous matrix. *Acta Mater.* **2004**, *52*, 5461–5469. [CrossRef]

47. Shi, X.; Cui, L.; Jiang, D.; Yu, C.; Guo, F.; Yu, M.; Ren, Y.; Liu, Y. Grain size effect on the R-phase transformation of nanocrystalline NiTi shape memory alloys. *J. Mater. Sci.* **2014**, *49*, 4643–4647. [CrossRef]
48. Wan, D.; Komvopoulos, K. Thickness effect on thermally induced phase transformations in sputtered titanium-nickel shape-memory films. *J. Mater. Res.* **2005**, *20*, 1606–1612. [CrossRef]
49. Chen, Y.; Schuh, C.A. Size effects in shape memory alloy microwires. *Acta Mater.* **2011**, *59*, 537–553. [CrossRef]

© 2019 by the authors. Licensee MDPI, Basel, Switzerland. This article is an open access article distributed under the terms and conditions of the Creative Commons Attribution (CC BY) license (http://creativecommons.org/licenses/by/4.0/).

Article

Optical and Superhydrophilic Characteristics of TiO$_2$ Coating with Subwavelength Surface Structure Consisting of Spherical Nanoparticle Aggregates

Yuki Kameya * and Hiroki Yabe

Department of Mechanical Engineering, Chiba Institute of Technology, Chiba 275-0016, Japan
* Correspondence: yuki.kameya@it-chiba.ac.jp or yuki.kameya.jp@gmail.com; Tel.: +81-47-478-0265

Received: 17 August 2019; Accepted: 24 August 2019; Published: 26 August 2019

Abstract: It is expected that the applications of photocatalytic coatings will continue to extend into many areas, so it is important to explore their potential for enhanced functionality and design flexibility. In this study, we investigated the effect of a subwavelength surface structure in a TiO$_2$ coating on its optical and superhydrophilic characteristics. Using submicron-scale spherical aggregates of TiO$_2$ nanoparticles, we fabricated a TiO$_2$ film with a subwavelength surface structure. Optical examination showed the enhanced transmittance of visible light compared to that of a plain surface. This was considered to be a result of a graded refractive index at the air–TiO$_2$ interface. The effect of the subwavelength surface structure on optical transmittance was also demonstrated by the numerical simulation of visible light propagation in which Maxwell's equations were solved using the finite-difference time-domain method. In addition, superhydrophilic behavior without ultraviolet light illumination was observed for the subwavelength-structure film via the measurement of the contact angle of a water drop. Furthermore, it was confirmed that the photocatalytic activity of the proposed film was comparable with that of a standard TiO$_2$ film. It was suggested that the control of the subwavelength surface structure of a TiO$_2$ film could be utilized to achieve novel properties of photocatalytic coatings.

Keywords: photocatalyst; TiO$_2$; nanoparticle; subwavelength surface structure; superhydrophilicity

---

1. Introduction

Titanium dioxide (TiO$_2$) has been widely used to decompose pollutants via photocatalytic reactions [1,2] and to control surface wettability for self-cleaning coatings [3]. Because TiO$_2$ has a high refractive index ($n \sim 2.6$ [4]) and no absorption band in the visible wavelengths, surface coatings of TiO$_2$ appear white. This intrinsic property limits the color design flexibility of TiO$_2$ coatings. Moreover, although TiO$_2$ is known to exhibit superhydrophilicity when illuminated with ultra-violet (UV) light, an external source of UV light is necessary to utilize its superhydrophilic function [3,5]. Therefore, it is desirable to develop TiO$_2$ films that provide enhanced transparency and superhydrophilicity without the need for UV-light irradiation.

To improve visible-light transmission at optical interfaces, one possible approach is to fabricate anti-reflection surface structures. A subwavelength surface pattern that enables an optical interface to have a graded refractive index which reduces light reflection [6]. Such surfaces are called moth-eye anti-reflective structures [7]. Considering this anti-reflection mechanism, it should be possible to increase the transparency of a TiO$_2$ film by fabricating a film with such subwavelength surface structures (i.e., structures having characteristic dimensions less than visible-light wavelengths).

Furthermore, the surface roughness of a hydrophilic material is known to enhance its apparent hydrophilicity [8]. Because TiO$_2$ is intrinsically hydrophilic, a textured surface on the TiO$_2$ film has the potential to provide increased hydrophilicity.

On the basis of the above considerations, it is expected to control the microscale surface structure of the TiO_2 film to achieve favorable optical and wetting characteristics. Recently, TiO_2 spherical nanoparticle aggregate (NSA) was developed [9]. The diameter of NSA is 100–200 nm, and, therefore, the top layer of a packed NSA film cannot be a plain surface but has an intrinsic subwavelength surface structure (Figure 1). Because such films consist of TiO_2 nanoparticles, they also have porous structures. A porous TiO_2 structure allows gas molecules to diffuse into the film, which results in the effective production of radicals before photo-excited carriers are lost [1]. Consequently, an NSA film is considered to show improved optical and wetting characteristics without the degradation of its photocatalytic function.

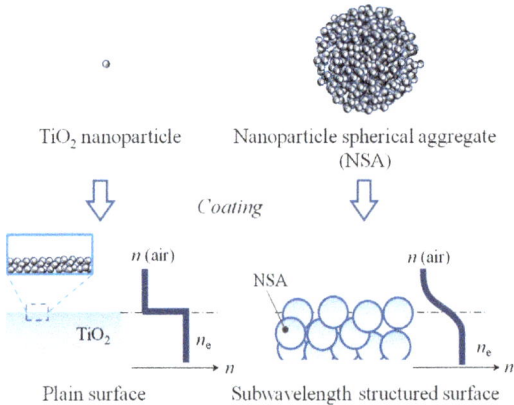

Figure 1. Illustration of a graded refractive index at the air–TiO_2 interface using a spherical nanoparticle aggregate (NSA) film.

In this work, we investigated the effect of the subwavelength surface structure of a TiO_2 coating on its optical and superhydrophilic characteristics. Using submicron-scale TiO_2 NSAs, we fabricated a TiO_2 film with a subwavelength surface structure. Visible-light transmittance was measured to examine the optical property resulting from the graded refractive index at the air–TiO_2 interface. The effect of the surface structure on the visible-light transmittance was also demonstrated by the numerical simulation of light propagation in which Maxwell's equations were solved using the finite-difference time-domain method. In addition, we performed water-drop contact-angle measurements to demonstrate superhydrophilic behavior without UV-light illumination. Finally, we confirmed the photocatalytic activity of an NSA film.

2. Experimental

2.1. TiO_2 Film Samples

We prepared TiO_2 aqueous dispersions using two kinds of TiO_2 particles: NSA (Ujiden Chemical, Kochi, Japan) and AEROXIDE® P25 (Evonik Industries, Essen, Germany). P25 is a fine powder of nanoparticles with mean diameter of about 21 nm. It was used as a standard TiO_2 material to evaluate the properties of NSA. To break large agglomerates in each dispersion, ultrasonication was performed using an UH-50 ultrasonic homogenizer (SMT, Kanagawa, Japan). A glass substrate was treated with Ar plasma to clean the surface and enhance its wettability. A specified amount of dispersion was dropped on the glass substrate and spontaneously spread over the surface. We managed to apply it to uniformly cover the substrate surface. After being dried at room temperature, samples were heated at 600 °C for 20 min to sinter the particles [10].

The prepared samples were observed using a scanning electron microscope (SEM) S-4700 (Hitachi, Tokyo, Japan). The glass plate caused surface charging during SEM observation; therefore, we coated the samples with osmium using a Neoc coater (Meiwafosis, Tokyo, Japan).

2.2. Optical Measurement

The optical transmittance of each sample was measured using a flame spectrometer with a halogen light source (Ocean Optics, Dunedin, FL, USA). A diffraction grating was used to split the light, and the intensity of each component was detected by a silicon CCD array. The measurement was performed for wavelengths in the range of 0.4–0.9 µm. The normal spectral transmittance was determined.

2.3. Water Contact Angle Measurement

To evaluate the wetting behavior, we measured the water contact angle at the surface. A water droplet (5 µL) was applied to the sample surface, and a side-view photo image was captured using a CMOS camera (3R-MSUSB401, 3R Solution, Fukuoka, Japan). Then, the obtained imaged was analyzed to determine the contact angle [11].

2.4. Photocatalytic Performance Evaluation

Methylene blue (MB) decomposition was used to evaluate the photocatalytic activity of prepared samples. We deposited an aqueous solution of MB on a TiO_2 film samples. We used a near-ultra-violet (NUV) lamp (FPL27BLB, Sankyo Denki, Kanagawa, Japan). The light intensity was about 4 mW/cm^2, which was measured using a UV light meter (UV-340C, Custom, Tokyo, Japan). Because MB has an optical absorption band in visible wavelengths, the change of optical transmittance was measured for three hours to evaluate the progress of MB decomposition.

3. Numerical Simulation

To investigate the effects of the surface subwavelength microstructure of a TiO_2 coating on the transmittance of visible light, we performed a numerical simulation of visible-light propagation through a textured surface. Maxwell's equations were solved using the finite-difference time-domain (FDTD) method [12]. We assumed optical properties and geometry to model the air–TiO_2 interface of the NSA film, as described below.

Because NSA consists of sintered nanoparticles, it has a fine porous structure. To evaluate the refractive index of the porous medium, we introduced the effective refractive index n_e, calculated from the Bruggeman effective medium approximation [13]:

$$f_1 \frac{(n_1^2 - n_e^2)}{(n_1^2 + 2n_e^2)} + f_2 \frac{(n_2^2 - n_e^2)}{(n_2^2 + 2n_e^2)} = 0 \tag{1}$$

where f is the volume fraction and the subscripts 1 and 2 refer to air and TiO_2, respectively, in the present case. For the substrates before sintering, we assumed that the porosity was 0.4 (random packing of spheres [1]), so $f_1 = 0.4$ and $f_2 = 0.6$. When we considered $n_1 = 1$ and $n_2 = 2.6$ [4], we obtained $n_e = 1.93$. Therefore, we used $n_e = 1.93$ as the effective refractive index of NSA in the present numerical simulation.

Then we modeled the subwavelength surface structure of the NSA film. The model geometry for the simulation is schematically shown in Figure 2a. Because of the spherical structure of each NSA, we assumed a convex surface pattern. The volume fraction of NSA at the air–TiO_2 interface varied along the surface-normal direction (i.e., the x direction); hence, a graded refractive index was achieved as illustrated in Figure 1. To examine the influence of the graded refractive index on the visible light transmission, we made several geometries, as shown in Figure 2b. We defined the aspect ratio of each convex structure as the height h divided by the width w. A parabolic function was used to create convex interface patterns with various aspect ratios. Because our NSA sample had a range of diameters, we used w = 0.1 and 0.2 µm, and the aspect ratio was varied from 0 to 0.5. Hence the height h of each convex structure is equal to the peak x value appearing at z = 0.05 and 0.1 µm for w = 0.1 and 0.2 µm, respsectively. The aspect ratio of 0 corresponds to a plain interface.

Because we used a periodic concave pattern to model the air–TiO$_2$ interface, the periodic boundary condition was used in the z direction (Figure 2a). For the x direction, we used the anisotropic perfectly matched layer (APML) absorbing boundary conditions [12]. The simulation region ($x \times z$) was 6 µm × 0.2 µm with the mesh size $\Delta x = \Delta z$ = 5 nm. A monochromatic plane wave of 0.55 µm wavelength, which is in the middle of visible range, was used to evaluate the transmittance of visible light. We investigated the influence of aspect ratio on the transmittance. As a solver, we used the commercial software OptiFDTD version 12 (Optiwave Systems) [14].

To confirm the validity of our simulation and subsequent data processing, we numerically calculated the normal transmittance of visible light (λ = 0.55 µm) regarding a plain TiO$_2$ layer and compared the result with the analytical solution of Maxwell's equations. The transmission coefficient τ at the air–TiO$_2$ interface is given as [15]:

$$\tau = \frac{4 n_1 n_e}{(n_1 + n_e)^2} \quad (2)$$

It should be noted that the transmittance T determined in the optical measurement has physical meaning different from the transmission coefficient τ; T includes the effect of multiple reflections at several interfaces in the sample, while τ is determined only by the phenomenon at the air–TiO$_2$ interface.

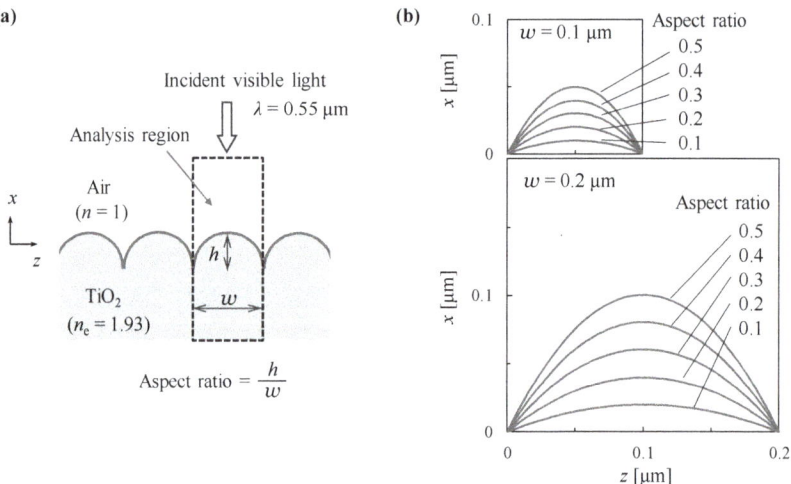

Figure 2. Numerical simulation model: (a) Analysis region. (b) Model geometries with various aspect ratios.

4. Results and Discussion

4.1. Visible-Light Transmittance

First, we visually examined the NSA film by comparing it with the P25 film. We prepared the dispersions of NSA and P25 with 30 wt % solid content, and each dispersion was coated onto black paper (Figure 3a). It is clear that the color change (i.e., black to white) observed in the P25-coated area was greater than that in the NSA-coated area. This result suggests that the visible light scattering by the NSA-coated surface was weaker than that from the P25-coated surface.

To confirm quantitatively the difference between the NSA and P25 films in terms of optical property, we measured the transmittance of TiO$_2$-coated samples in the visible wavelength range. The dispersion of NSA or P25 with 5 wt % solid content was used to make film on a glass substrate for each sample. Then, the normal spectral transmittance of each sample was measured. The results are shown in Figure 3b. The transmittance of the NSA film is higher than that of P25 in the visible wavelengths. For example, the result at the wavelength of 0.55 µm showed T = 62% and 51% for NSA and P25, respectively, so the difference between these films was significant. The obtained results are in accordance with the above-mentioned visual examination using black paper.

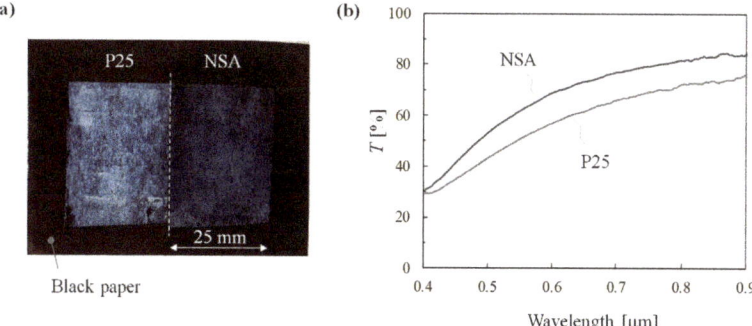

Figure 3. (a) Photograph of a coating surface on black paper. (b) Spectral transmittance in the visible wavelength range.

4.2. SEM Observation

The TiO$_2$ film samples used in the optical transmission measurement were observed using SEM. The obtained top-view images are shown in Figure 4. The P25 film had no unique geometrical features (Figure 4a), and the surface roughness that existed was possibly due to the agglomeration of particles during the process of drying the dispersion [16]. Concerning the NSA film (Figure 4b), the spherical shape of NSA remained even after the sintering process. As expected, the surface structure of each film reflected the basic-unit size of particles, i.e., the primary-particle diameter of about 21 nm for P25 and the aggregate diameter of 100–200 nm for NSA. We confirmed a remarkable difference in the qualitative characteristics of surface structures between the two samples. Hence, we proceeded to the numerical simulation of visible light transmission to explain the observed difference in the transmittance.

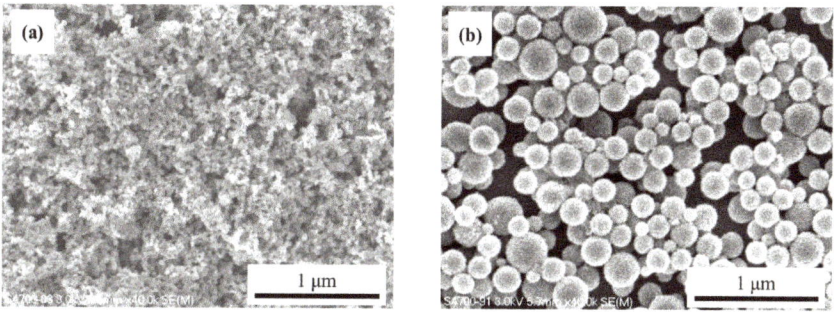

Figure 4. Top-view SEM images: (a) P25 and (b) NSA.

4.3. Numerical Simulation of Visible-Light Transmission

The transmission coefficient τ at the air–TiO$_2$ interface was numerically calculated for each geometry described in Section 2 (Figure 2b). The simulation results are summarized in Figure 5.

The transmission coefficients τ are plotted against the aspect ratio of a concave surface structure. The circle and triangle symbols show results of w = 0.1 and 0.2 µm, respectively.

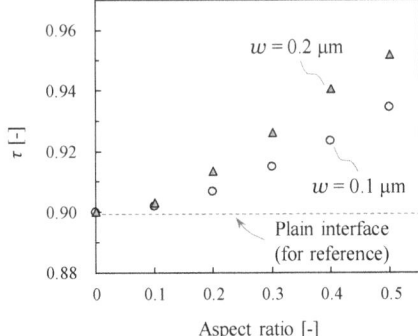

Figure 5. Simulated transmission coefficient as a function of aspect ratio (circles: w = 0.1 µm, and triangles: w = 0.2 µm). A value corresponding to the transmission coefficient of a plain interface is indicated by the horizontal dashed line for reference.

The case in which the aspect ratio was zero corresponds to a plain interface, so the predicted τ values for two results (w = 0.1 and 0.2 µm) should be the same value. Using Equation (2), the transmission coefficient obtained for a plain interface was 0.899. Therefore, the validity of our numerical simulation model and subsequent data processing was confirmed.

As the aspect ratio increased, the transmission coefficient showed higher values. This trend is in accordance with our expectation because the effect of a graded refractive index at the air–TiO$_2$ interface is emphasized at higher aspect ratios.

The difference between the cases of w = 0.1 and 0.2 µm can be also interpreted as indicating spatial variation of the refractive index. The volume fraction of a convex structure for the aspect ratio of 0.5 is plotted against the x coordinate (i.e., surface normal direction) in Figure 6. We found that the gradient of the curve for w = 0.2 µm was lower than that for w = 0.1 µm. The lower gradient of the curve means spatially broader distribution of the refractive index, which results in transmission enhancement.

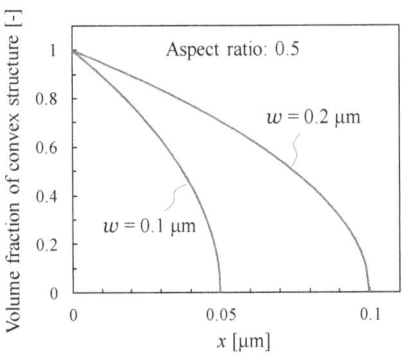

Figure 6. Volume fraction of a convex structure with surface normal (x-axis) direction.

On the basis of these simulation results, we expect an NSA film with a subwavelength surface structure should show improved transparency. Considering the increasing transmission coefficient with the aspect ratio, there is potential for improvement by modifying the surface structure of the NSA film.

4.4. Wetting

Wettability is an important property of coatings intended to achieve a self-cleaning surface. To evaluate the water wettability of our TiO$_2$ samples, we performed contact angle measurement. The photo images taken to determine the contact angle θ_c are shown in Figure 7. To emphasize the change in the contact angle for the TiO$_2$ coating, we also used a glass slide that was exposed to air and not coated with TiO$_2$. The uncoated glass had $\theta_c = 34°$, which is not an ideal result for a clean glass surface but a practical result for a glass surface exposed to airborne contaminants (Figure 7a). For the P25-coated surface, a drop of water readily spread over the surface and eventually showed $\theta_c < 10°$, which is a characteristic of superhydrophilic surfaces (Figure 7b). The NSA-coated surface also exhibited superhydrophilic behavior (Figure 7c). Additionally, a sessile-drop technique is often used to distinguish advancing and receding contact angles [17,18]. Because a water drop was simply deposited on each sample in the present experiments, the results are close to advancing contact angles.

Even though we did not use UV light irradiation in these experiments, we observed superhydrophilic behavior for both the P25 and NSA coatings. We consider that the surface roughness influenced the small contact angle. The TiO$_2$ coatings, not only with NSA, but also with P25, exhibited surface roughness. Even though their scales and features were totally different, it is considered that the roughness of both films was sufficient to enhance their water wetting property; this consideration can be supported by the observation indicating that nanoscale roughness affects surface wetting [19].

Because it is difficult to accurately measure very small contact angles ($\theta_c < 10°$), we were not able to determine the difference in wettability of the P25 and NSA coatings in the present experiments. To quantitatively investigate the difference in the degree of superhydrophilicity, a special experimental setup is needed [20]. Future work is necessary to optimize the surface microstructure proposed here.

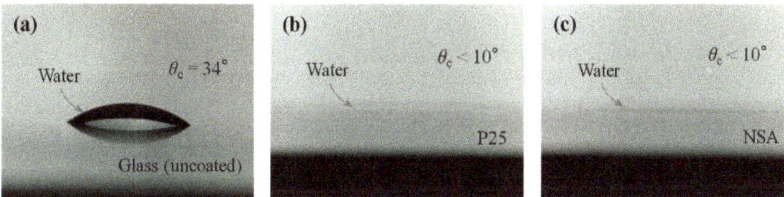

Figure 7. Contact angle measurement: (**a**) Uncoated glass, (**b**) P25 film, and (**c**) NSA film.

4.5. Photocatalytic Decomposition of MB

To demonstrate the photocatalytic performance of the NSA-coated film, we conducted MB decomposition experiments. After depositing MB on the NSA and P25 films, we measured their visible-light transmittance to evaluate the amount of MB on the samples. Then, we started to irradiate them with UV light to enhance the photocatalytic decomposition of MB. We measured the visible-light transmittance every hour.

The results are shown in Figure 8. At the beginning (the time was 0 h), the transmittance of the NSA film was higher than that of the P25 film because of the difference in their clean conditions (shown in Figure 3b). As the reaction proceeded, the transmittance of each sample increased due to decomposition of light-absorbing MB. The temporal change of transmittance indicates the photocatalytic reaction rate of MB decomposition. Because the slope of transmittance was similar for each sample, we can assume there was no remarkable difference in the MB decomposition rates. In this way, we confirmed that the NSA film had photocatalytic performance comparable to that of a typical film with the TiO$_2$ coating using P25. Therefore, we expect to be able to utilize the useful features of NSA films without reducing their photocatalytic activity.

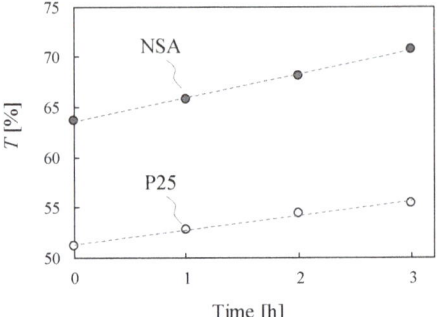

Figure 8. Transmittance variation with time during photocatalytic methylene blue (MB) decomposition.

5. Conclusions

We investigated the effect of a subwavelength surface structure of a TiO$_2$ coating on its optical and superhydrophilic characteristics. Using submicron-scale TiO$_2$ NSA, we fabricated a TiO$_2$ film with a subwavelength surface structure. Optical examination showed an enhanced transmittance of visible light compared to that of a plain surface, which was considered to be a result of a graded refractive index at the air–TiO$_2$ interface. The numerical simulation supported the improved transparency. In addition, superhydrophilic behavior without ultraviolet-light illumination was observed for the subwavelength-structured film via the contact angle measurement of a water drop. It is suggested that we can utilize the favorable features of an NSA film, such as greater visible-light transmission and superhydrophilicity, without degrading its photocatalytic performance. Using a subwavelength surface structure on a photocatalytic film appears to influence its useful features, so it is hoped to further investigate the potential for surface design of photocatalytic coatings.

Author Contributions: Conceptualization, Y.K.; Methodology, Y.K.; Formal Analysis, Y.K. and H.Y.; Investigation, Y.K. and H.Y.; Writing—Review and Editing, Y.K.

Funding: This research received no external funding.

Acknowledgments: The authors thank Tomohiro Okazoe of Ujiden Chemical Industry for supplying the TiO$_2$ materials and useful suggestions. A part of the experiments and numerical simulation were conducted by Takahiro Yamada and Daiki Sato as an undergraduate research program of Chiba Institute of Technology.

Conflicts of Interest: The authors declare no conflicts of interest.

References

1. Kameya, Y.; Torii, K.; Hirai, S.; Kaviany, M. Photocatalytic soot oxidation on TiO$_2$ microstructured substrate. *Chem. Eng. J.* **2017**, *327*, 831–837. [CrossRef]
2. Truppi, A.; Luna, M.; Petronella, F.; Falcicchio, A.; Giannini, C.; Comparelli, R.; Mosquera, M.J. Photocatalytic activity of TiO$_2$/AuNRs-SiO$_2$ nanocomposites applied to building materials. *Coatings* **2018**, *8*, 296. [CrossRef]
3. Banerjee, S.; Dionysiou, D.D.; Pillai, S.C. Self-cleaning applications of TiO$_2$ by photo-induced hydrophilicity and photocatalysis. *Appl. Catal. B Environ.* **2015**, *176*, 396–428. [CrossRef]
4. Chen, T.L.; Hirose, Y.; Hitosugi, T.; Hasegawa, T. One unit-cell seed layer induced epitaxial growth of heavily nitrogen doped anatase TiO$_2$ films. *J. Phys. D Appl. Phys.* **2008**, *41*, 062005. [CrossRef]
5. Wang, R.; Hashimoto, K.; Fujishima, A.; Chikuni, M.; Kojima, E.; Kitamura, A.; Shimohigoshi, M.; Watanabe, T. Light-induced amphiphilic surfaces. *Nature* **1997**, *388*, 431–432. [CrossRef]
6. Lin, H.; Ouyang, M.; Chen, B.; Zhu, Q.; Wu, J.; Lou, N.; Dong, L.; Wang, Z.; Fu, Y. Design and fabrication of moth-eye subwavelength structure with a waist on silicon for broadband and wide-angle anti-reflection property. *Coatings* **2018**, *8*, 360. [CrossRef]
7. Loh, J.Y.Y.; Kherani, N. Design of nano-porous multilayer antireflective coatings. *Coatings* **2017**, *7*, 134. [CrossRef]

8. Drelich, J.; Chibowski, E.; Meng, D.D.; Terpilowski, K. Hydrophilic and superhydrophilic surfaces and materials. *Soft Matter* **2011**, *7*, 9804–9828. [CrossRef]
9. Wang, P.; Kobiro, K. Ultimately simple one-pot synthesis of spherical mesoporous TiO_2 nanoparticles in supercritical methanol. *Chem. Lett.* **2012**, *41*, 264–266. [CrossRef]
10. Kameya, Y.; Yamaki, H.; Ono, R.; Motosuke, M. Fabrication of micropillar TiO_2 photocatalyst arrays using nanoparticle-microprinting method. *Mater. Lett.* **2016**, *175*, 262–265. [CrossRef]
11. Kameya, Y. Wettability modification of polydimethylsiloxane surface by fabricating micropillar and microhole arrays. *Mater. Lett.* **2017**, *196*, 320–323. [CrossRef]
12. Taflove, A.; Hagness, S.C. *Computational Electrodynamics: The Finite-Difference Time-Domain Method*, 3rd ed.; Artech House: Norwood, MA, USA, 2005.
13. Chattopadhyay, S.; Huang, Y.F.; Jen, Y.J.; Ganguly, A.; Chen, K.H.; Chen, L.C. Anti-reflecting and photonic nanostructures. *Mater. Sci. Eng. R* **2010**, *69*, 1–35. [CrossRef]
14. OptiFDTD. Available online: https://optiwave.com/optifdtd-overview/ (accessed on 14 August 2019).
15. Griffiths, D.J. *Introduction to Electrodynamics*, 4th ed.; Cambridge University Press: Cambridge, UK, 2017.
16. Kameya, Y. Kinetic Monte Carlo simulation of nanoparticle film formation via nanocolloid drying. *J. Nanopart. Res.* **2017**, *19*, 214. [CrossRef]
17. Drelich, J. Guidelines to measurements of reproducible contact angles using a sessile-drop technique. *Surf. Innov.* **2013**, *1*, 248–254. [CrossRef]
18. Marmur, A.; Volpe, C.D.; Siboni, S.; Amirfazli, A.; Drelich, J.W. Contact angles and wettability: towards common and accurate terminology. *Surf. Innov.* **2017**, *5*, 3–8. [CrossRef]
19. Wang, H. From contact line structures to wetting dynamics. *Langmuir* **2019**, *35*, 10233–10245. [CrossRef] [PubMed]
20. Allred, T.P.; Weibel, J.A.; Garimella, S.V. A wettability metric for characterization of capillary flow on textured superhydrophilic surfaces. *Langmuir* **2017**, *33*, 7847–7853. [CrossRef] [PubMed]

© 2019 by the authors. Licensee MDPI, Basel, Switzerland. This article is an open access article distributed under the terms and conditions of the Creative Commons Attribution (CC BY) license (http://creativecommons.org/licenses/by/4.0/).

Article

Spectrophotometric Characterization of Thin Copper and Gold Films Prepared by Electron Beam Evaporation: Thickness Dependence of the Drude Damping Parameter

Olaf Stenzel [1,*], Steffen Wilbrandt [1], Sven Stempfhuber [1,2], Dieter Gäbler [1] and Sabrina-Jasmin Wolleb [1]

[1] Fraunhofer Institute of Applied Optics and Precision Engineering IOF, 07745 Jena, Albert-Einstein Str. 7, Germany; Steffen.Wilbrandt@iof.fraunhofer.de (S.W.); Sven.Stempfhuber@iof.fraunhofer.de (S.S.); Dieter.Gaebler@iof.fraunhofer.de (D.G.); Sabrina-Jasmin.Wolleb@iof.fraunhofer.de (S.-J.W.)
[2] Abbe School of Photonics, Friedrich-Schiller-University Jena, Albert-Einstein-Str. 6, 07745 Jena, Germany
* Correspondence: Olaf.Stenzel@iof.fraunhofer.de; Tel.: +49-3641-807-348

Received: 19 February 2019; Accepted: 5 March 2019; Published: 9 March 2019

Abstract: Copper and gold films with thicknesses between approximately 10 and 60 nm have been prepared by electron beam evaporation and characterized by spectrophotometry from the near infrared up to the near ultraviolet spectral regions. From near normal incidence transmission and reflection spectra, dispersion of optical constants have been determined by means of spectra fits utilizing a merger of the Drude model and the beta-distributed oscillator model. All spectra could be fitted in the full spectral region with a total of seven dispersion parameters. The obtained Drude damping parameters shows a clear trend to increase with decreasing film thickness. This behavior is discussed in the context of additional non-optical characterization results and turned out to be consistent with a simple mean-free path theory.

Keywords: copper; gold; ultrathin metal films; optical constants; thickness dependence

---

1. Introduction

In thin film characterization and design practice, numerous dispersion models exist that may be used for reliable modelling of the optical constants of dielectric films, regardless of whether they are transparent or absorbing. Examples of suitable models are provided by the oscillator model [1,2], Tauc-Lorentz [3] and Cody Lorentz [4] models, Gaussian-like broadened oscillators [5,6], or the β_do model [7]. On the other hand, there is still uncertainty among optical coating practitioners concerning a realistic treatment of the optical constants of metal films, and in particular of ultrathin metal films. This is irritating, as thin metal films are widely used in optical transmission filters [8] and architectural glass coatings [9], just to give two examples.

In fact, it was already shown in 1984 in a famous round robin experiment [10] that the determination of thickness and optical constants of two (ultra-)thin rhodium films (thicknesses around 15 and 30 nm) turned out to be much more challenging than the analysis of sufficiently thick dielectric (in this case Sc_2O_3) coatings. Differing results have been obtained by means of different optical characterization techniques applied by the participating research groups. Since 1984 much work was done, nevertheless the uncertainty among practitioners in managing ultrathin metal film properties cannot be eliminated.

In May 2018 at the SPIE conference "Advances in Optical Coatings VI" (SPIE conference 10691), some of speakers' time slots turned out to be unexpectedly vacated because of speakers missing due to airline strikes. These time slots were filled ad hoc with a standby discussion on current problems in the theoretical description of ultrathin metal film optical properties. Here, the authors of the present study developed their point of view that it would be useful to adopt the mean free path theory successfully applied in solid state physics [11,12] and cluster physics [13] to the modelling of thin metal film optical constants. It is the purpose of this paper to re-examine this idea and to demonstrate experimental examples on the use of this treatment.

The general idea is rather simple. One may start from the classical Drude function [2] in order to model the dielectric response of the free carrier fraction in metals. The bound electron response may be modeled in terms of some derivative of the multioscillator model [1–7]. For smooth closed films, such a treatment leads to satisfactory spectra fits corresponding to thickness-independent optical constants, as long as the film thickness is large enough (as a rule of thumb, the films should be thicker than approximately 50 nm).

However, when the film thickness d becomes smaller than approximately twice the mean free path of the conduction electrons in the bulk, the picture may change. Then, collisions of conduction electrons with the film surface lead to a decrease in the mean free path, which may result in an increase in the damping parameter Γ_D of the Drude function. In smooth closed thin films, this effect is expressed in terms of the parameter $\frac{v_F \tau_b}{d}$ [11,12]. Here, v_F is the Fermi velocity, and τ_b the average time between two collisions suffered by a free charge carrier in the bulk.

Note that an analogous approach is in use in small metal cluster optics [13]. Here the parameter is $A \frac{v_F \tau_b}{R}$ with R—cluster radius, and A is a constant in the range of one.

In the following, we will demonstrate the validity of such an approach to the optical properties of thin copper and gold films.

2. Theory

2.1. Dispersion Model

In order to account for both the free and bound electrons fractions in a metal film, we make use of the following writing of the metal's dielectric function ε (n–refractive index; k–extinction coefficient):

$$\varepsilon = 1 + \chi_{\text{free}} + \chi_{\text{bound}} = (n + ik)^2 \quad (1)$$

where χ_{free} is the susceptibility characterizing the free electron fraction (corresponding to intraband transitions), and χ_{bound} describes the contribution of interband transitions, i.e., the response of the bound electrons.

For the free electrons, we will use the classical Drude function [2]:

$$\chi_{\text{free}} = -\left(\frac{v_p^2}{v^2 + 2iv\Gamma_D} \right) \quad (2)$$

Here v is the wavenumber (e.g., the reciprocal value of the vacuum wavelength), and v_p the free electrons plasma frequency in wavenumber units:

$$v_p = \frac{\omega_p}{2\pi c} = \frac{E_p}{2\pi \hbar c} \quad (3)$$

where (c is velocity of light in the vacuum, E_p is bulk plasmon energy, and ω_p is bulk plasma (angular) frequency) Γ_D is the Drude damping parameter, again in wavenumber units. For the bulk material we have:

$$\Gamma_{D,b} = \frac{1}{4\pi c \tau_b} \quad (4)$$

Particularly, for the "perfect" bulk crystal, Equation (4) reads as:

$$\Gamma_{D,b0} = \frac{1}{4\pi c \tau_{b0}} \quad (5)$$

Theoretical values on the bulk collision time τ_{b0} in the perfect crystal and the Fermi velocity v_F are given in Table 1 together with selected other parameters known for bulk copper and gold from the literature.

Table 1. Theoretical data (E_{a0}: photon energy corresponding to the onset of absorption features caused by interband transitions in the optical spectra of the perfect crystal, ν_{a0} is the corresponding onset wavenumber).

Metal	τ_{b0}/fs [14]	$v_F\tau_{b0}$/nm [14]	ϱ/g cm^{-3}	E_p/eV	E_{a0}/eV [15]	$\Gamma_{D,b0}$/cm^{-1} (Equation (5))	ν_p/cm^{-1} (Equation (3))	ν_{a0}/cm^{-1}
Cu	36	39.9	8.92	9.3 [16]	2.09	74	75,004	16,856
Au	27.3	37.7	19.32	8.83 [15]	2.35	97	71,214	18,953

Different sets of copper and gold Drude parameters (Table 2) are obtained or used in studies published in the last decades. The reported data scatter strongly, but are of the same order of magnitude as the values presented in Table 1.

Table 2. Survey of literature data on gold and copper Drude parameters. Numerical values have been adopted to the writing of the Drude function given in Equation (2).

Metal	Reference	ν_p/cm^{-1}	Γ_D/cm^{-1}
Cu	[17]	59600	36
	[18]	70660	385
Au	[17]	72800	112
	[18]	71710	285
	[19]	72200	279
	[20]	72590	282
	[21]	69000	74

It is worth noting that in reference [20], Mie resonances in nanosized gold clusters with different diameters have been studied, and the increase in the Drude damping parameter with decreasing cluster diameter could be directly observed. In the smallest clusters with diameters of 5 nm, Γ_D reached values up to approximately 650 cm^{-1}.

According to Equation (1), a realistic description of the metal's optical properties must also contain the response of the bound electrons. In this study, in order to quantify χ_{bound}, we make use of the beta-distributed oscillator (β_do) dispersion model [7]. In the β_do model, the susceptibility χ_{bound} is given by:

$$\chi_{bound}(\nu) = \frac{J}{\pi} \frac{\sum\limits_{s=1}^{N} w_s \left(\frac{1}{\nu_s - \nu - i\Gamma} + \frac{1}{\nu_s + \nu + i\Gamma} \right)}{\sum\limits_{s=1}^{N} w_s}; \qquad (6)$$

$$w_s = \left(\frac{s}{N+1}\right)^{\alpha-1} \left(\frac{N+1-s}{N+1}\right)^{\beta-1}; \quad s = 1, 2, 3, \ldots, N$$

$$\nu_s = \nu_a + \frac{\nu_b - \nu_a}{N+1} s$$

In fact, Equation (6) describes inhomogeneous broadening of an absorption feature with a shape described in terms of the beta-function, with exponents defined by α and β. N is the number of equidistant Lorentzian oscillators (with homogeneous linewidth Γ) considered in numerical modelling, while ν_a and ν_b mark onset and cutoff wavenumbers of the inhomogeneously broadened absorption feature. J is an intensity factor proportional to the oscillator strength. The mentioned parameters form the set of β_do dispersion model parameters. The optical constants of the films, as well as the Drude and β_do dispersion parameters, may then be obtained from fitting experimental transmission and reflection spectra of the real films in terms of Equations (1), (2), and (6). The optical constants in this study have been determined in this way.

2.2. Film Thickness Estimation

In order to determine the metal film thickness, three different approaches have found an application:

- Thicknesses have been determined from fitting experimental transmission and reflection spectra in terms of Equations (1), (2), and (6), taking the film thickness as a further fitting parameter.
- Thicknesses have been determined from X-ray reflection (XRR) analysis.
- Thicknesses have been estimated from measured transmission and reflection spectra without assuming a specific dispersion law, but on the basis of a sum-rule-based theoretical approach using potential absorptance [22,23].

In point iii, thickness determination was achieved by first calculating the experimental value Ω_{exp} defined by

$$\Omega_{exp} \equiv \frac{\int_{357nm}^{2000nm} \left\{1 - \frac{T_{exp}(\lambda)}{[1-R_{exp}(\lambda)]}\right\} \frac{d\lambda}{\lambda^2}}{\int_{357nm}^{2000nm} \frac{d\lambda}{\lambda^2}} \tag{7}$$

for each sample from the measured near normal incidence transmission (T) and reflection (R) spectra. That value was compared to a theoretical value calculated based on tabulated optical constants [24] as a function of film thickness d:

$$\Omega_{calc}(d) \equiv \frac{\int_{357nm}^{2000nm} \left\{1 - \frac{T_{calc}(\lambda,d)}{[1-R_{calc}(\lambda,d)]}\right\} \frac{d\lambda}{\lambda^2}}{\int_{357nm}^{2000nm} \frac{d\lambda}{\lambda^2}} \tag{8}$$

Then, d was estimated from setting:

$$\Omega_{calc}(d) = \Omega_{exp} \tag{9}$$

As it will be seen later in Section 4, the film thicknesses obtained by these three methods are in good agreement, except metal films with a thickness of more than 50 nm.

2.3. Mean Free Path Effects

As has already been shown in earlier work [11,12], in ultrathin metal or semiconductor films, a drop in static electric conductivity occurs in ultrathin metal films when the film thickness becomes smaller than a value of approximately $2v_F\tau_b$ (for gold and copper, according to Table 1, this thickness corresponds to approximately 75 nm). The point is that in ultrathin films, collisions between charge carriers and the film surface may dominate over bulk collision effects, and this may lead to a significant reduction in the average collision time. According to the classical Drude theory, this will be accompanied by a decrease in the static electric conductivity, and an increase in the Drude damping parameter according to Equation (4).

As is shown in an earlier study [11], the simplest model treatment of this reduction in the mean free path of the charge carriers results in a thickness-dependent average collision time given by:

$$\tau(d) = \frac{\tau_b}{\left[1 + 2\frac{(1-p_{spec})l_{free}}{d}\right]} \tag{10}$$

Here, p_{spec} is the relative amount of charge carriers that is specularly reflected at the film surface, and l_{free} the mean free path in the bulk material. Note that in this approach, only charge carriers reflected diffusely from the surface contribute to the mentioned thickness dependence. When setting $l_{free} = v_F \tau_b$ [11,12] and making use of Equation (4), we obtain a manageable expression for the thickness-dependent Drude damping parameters in thin films according to:

$$\Gamma_D(d) = \Gamma_{D,b} \left[1 + \frac{2(1 - p_{spec}) v_F \tau_b}{d} \right] \tag{11}$$

This expression will be in the basis of our discussion of thickness-dependent Drude parameters in the ultrathin copper and gold films.

3. Experimental

3.1. Layer Deposition

The metal layers have been prepared in a Bühler Syrus pro 1110 coating machine (Alzenau, Germany) by e-beam evaporation using a tungsten liner. The Au-material was Au 5N+ from SAXONIA Technical Materials (Hanau, Germany) and the Cu material 5N grade supplied from Umicore Company (Balzers, Liechtenstein). Fused silica Q1 as well as silicon wafers have been used as substrates. No substrate preconditioning by plasma etching has been applied, and no adhesion layers have been deposited. The evaporation geometry was optimized with respect to the specific demands of metal coating, so the distance from evaporation surface to the substrate was only 32 cm, while the substrates were held at fixed positions without any rotation of the substrate holder. The pressure was 4.7×10^{-7} mbar prior to deposition start. The substrate temperature during evaporation was about 30 °C. The evaporation rate of 0.5 nm/s and the thicknesses have been controlled by quartz crystal monitoring.

3.2. Layer Characterization

3.2.1. Spectrophotometry

T- and R- spectra in the range of 350–2000 nm of all samples have been measured at near normal incidence in a Perkin Elmer Lambda 950 scanning spectrophotometer (Rodgau, Germany) equipped with absolute T- and R measurement attachments. From these spectra, film thickness d as well as optical constants n and k have been deduced from spectra fits in terms of Equations (1), (2), and (6), using a Matlab environment. In all spectra fits, the parameter N in Equation (6) was set to $N = 1000$. Generally, spectra from samples deposited onto both types of substrates (silicon and fused silica) have been included into the fitting procedure. In more detail, the fitting procedure is described in a previous study [7].

3.2.2. X-ray Reflection XRR

All samples were characterized with grazing incidence X-ray reflectivity (XRR) using a Bruker (Karlsruhe, Germany) D5005 diffractometer operated with Cu-Kα radiation (λ = 0.154 nm) in symmetrical θ–2θ geometry. The film thickness, material density, and surface roughness were extracted from simulation of the reflective curves by the commercial program "Leptos" (version 7.8).

3.2.3. Scanning Electron Microscopy SEM

The layer surfaces were investigated using a Carl Zeiss Sigma scanning electron microscope (SEM, Jena, Germany). The applied acceleration voltage was 5 kV and an InLens-Detector was used to detect the secondary electrons. To get a good overview, images were made in different magnifications. The presented results were made with a magnification of 80.00 K×.

4. Results

Concerning non-optical properties of the metal layers, we start with a short survey of the XRR and SEM results. For all copper layers on fused silica, an identical rms surface roughness of 1.0 nm has been found by XRR. The density values were scattered in the region (8.6 ± 0.2) g·cm^{-3} without a clear thickness dependence. Concerning the gold layers, the rms surface roughness data were scattered in the region (1.1 ± 0.2) nm, while the density was about (19.1 ± 0.1) g·cm^{-3}. All films have approximately the same surface roughness around 1 nm rms, while the average density turned out to be marginally smaller than typical bulk densities (96.5% for Cu, and 98.9% for gold—compare with Table 1).

In qualitative agreement with the rather thickness-independent surface roughness, all SEM pictures of copper films on fused silica show practically identical granular surface morphologies with grain sizes in the region of 20 nm (Figure 1a–c). The corresponding pictures of gold surfaces (Figure 1d–f) show different behavior. Again, granular areas are observed, but the dominating features are given by large material slabs with diameters up to several 100 nm that appear rather smooth. Furthermore, these material slabs show a trend of growing in size with increasing layer thickness. It may be expected that the large gold slabs result in optical properties of gold films that are much closer to the "perfect" bulk references, as would be expected in the case of copper films.

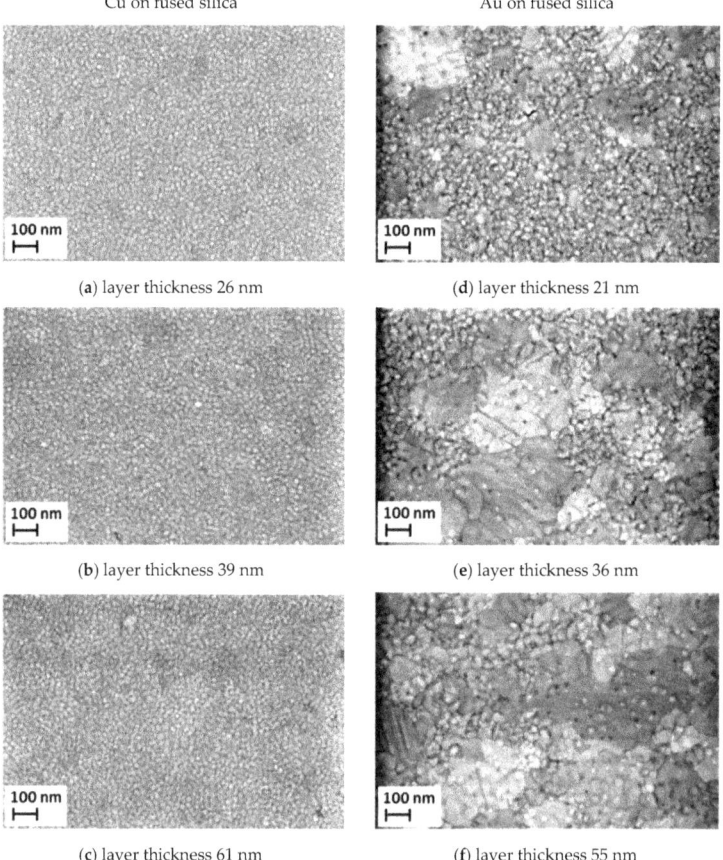

Figure 1. Selected SEM images of the metal films surface morphology. The mentioned thicknesses correspond to those obtained from the spectra fits.

Concerning the optical properties, let us first mention that all spectra could be well fitted by means of the chosen dispersion approach in Equations (1), (2), and (6), assuming α = β, such that a total of only seven independent dispersion parameters (two Drude parameters and five β_do parameters) were necessary to model the optical constants. The film thicknesses of the films on fused silica and silicon appear as additional parameters for the spectra fits. Figure 2 demonstrates the achieved fit quality for a gold (Figure 2a) a copper (Figure 2b) sample. The chosen dispersion approach obviously results in excellent spectra fits.

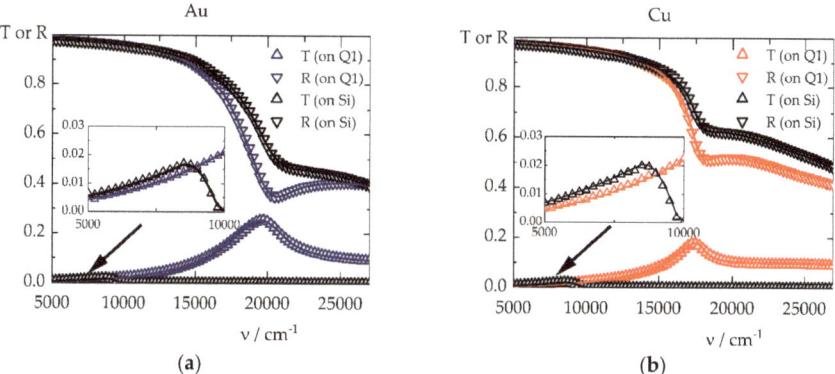

Figure 2. (**a**): experimental spectra (triangles) and fit (lines) of a gold film on Q1 (navy) and Si (black). On Q1, a thickness of 36.9 nm was found, on Si, the thickness was 41.1 nm; (**b**): experimental spectra (triangles) and fit (lines) of a copper film on Q1 (navy) and Si (black). On Q1, a thickness of 39.3 nm was found, on Si, the thickness was 42.9 nm.

Figure 3 shows the obtained optical constants for all samples. From the figures, we recognize the typical dispersion of metal optical constants in the spectral region from the near infrared to the visible. The comparison with literature data from an earlier study [24] (green circles) confirms the physical relevance of the obtained data. Generally, thinner layers tend to have slightly larger refractive indices in the near infrared, and slightly smaller extinction coefficients. The corresponding dispersion parameters are collected in the Tables 3 and 4.

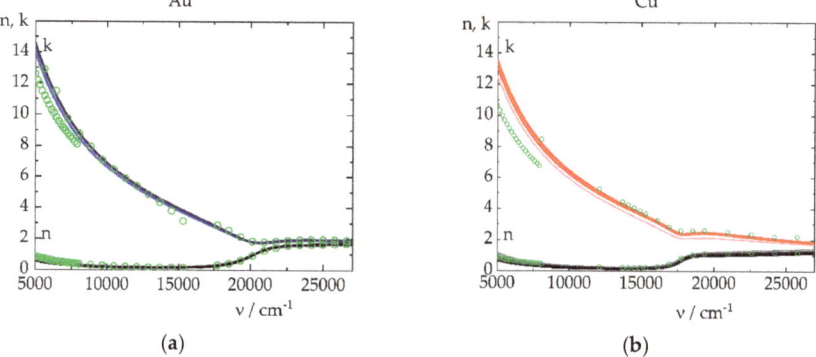

Figure 3. Optical constants of the gold (**a**) and copper (**b**) films. Curves that appear darker in color correspond to thicker films. Green circles represent literature data [24].

Table 3. Thickness and dispersion parameters of the gold films. Thickness data correspond to the samples on fused silica (Q1).

Fit	XRR	d_{Q1}/nm Equation (9)	ν_a/cm^{-1}	ν_b/cm^{-1}	J/cm^{-1}	Γ/cm^{-1}	$\alpha = \beta$	ν_p/cm^{-1}	Γ_D/cm^{-1}
20.7	19.0	20.0	20,740	83,340	399,370	2010	0.85	71,530	270
36.3	34.8	34.0	20,730	86,610	439,750	1760	0.87	74,000	200
54.9	53.6	53.2	20,730	87,740	445,170	1720	0.87	74,260	190
	$\delta f \rightarrow$		0.0004	0.05	0.11	0.16	0.02	0.04	0.36

Table 4. Thickness and dispersion parameters of the copper films. Thickness data correspond to the samples on fused silica (Q1).

Fit	XRR	d_{Q1}/nm Equation (9)	ν_a/cm^{-1}	ν_b/cm^{-1}	J/cm^{-1}	Γ/cm^{-1}	$\alpha = \beta$	ν_p/cm^{-1}	Γ_D/cm^{-1}
12.3	10.3	11.8	17,770	81,870	312,610	1020	0.91	64,400	460
20.1	17.0	18.4	17,850	86,900	326,510	820	0.88	68,310	360
26.2	21.7	23.9	17,840	86,790	325,930	770	0.88	69,090	340
39.3	34.3	35.3	17,840	86,350	310,050	720	0.87	69,270	280
61.3	50.0	52.7	17,820	85,820	274,490	710	0.87	66,450	240
	$\delta f \rightarrow$		0.005	0.06	0.17	0.38	0.05	0.07	0.64

In order to identify relevant thickness-dependent trends in the parameters given in Tables 3 and 4, for each dispersion parameter f, a relative dynamic range δf has been calculated as the difference between its maximum and minimum values in the corresponding column of the table, divided by the average of all data in the column. Hence, δf is given by:

$$\left[\frac{f_{max} - f_{min}}{\langle f \rangle}\right]\bigg|_{column} \equiv \delta f \qquad (12)$$

The spectra fit for the thickest copper layer was numerically instable, most probably because of the strong damping within the layer, indicated by the weak transmission signal. As a result we recognize the large mismatch between the thickness values obtained by the different methods. In Table 4, a set of dispersion parameters is included as obtained from a fit attempt when the difference $\nu_b - \nu_a$ was kept close to what is known from the bulk material. Table 5 shows average values and standard deviations of the fitting parameters obtained from different fit attempts of the spectra of the problematic thickest copper film. Fortunately, the obtained damping parameters that are in the focus of our study scatter with a relative standard deviation of no more than 12%. Thus, regardless of the principal numerical instability of the spectra fits, the data range for the damping parameters reported in Table 5 does not violate the thickness-dependent trends, which follow from Table 4. Moreover, all dispersion data presented in Table 4 for this copper sample fall into the ranges presented in Table 5. Therefore, we assume that the data of the thickest film, as included in Table 4, are consistent with the spectra and are physically meaningful, regardless of the numerical problems with that fit.

Table 5. Average values and standard deviations of the parameters of the thickest copper layer, as obtained from different fitting attempts.

d_{Q1}/nm	ν_a/cm^{-1}	ν_b/cm^{-1}	J/cm^{-1}
59 ± 7	17,800 ± 40	87,000 ± 15,000	330,000 ± 130,000
Γ/cm^{-1}	$\alpha = \beta$	ν_p/cm^{-1}	Γ_D/cm^{-1}
680 ± 40	0.89 ± 0.03	69,000 ± 7,000	250 ± 30

A comparison of the δf values from Tables 3 and 4 shows that in both metals, strongest relative changes from sample to sample (highest δf) are observed for the Drude damping parameter Γ_D, followed by the homogeneous linewidth of the β_do oscillators Γ, which also essentially represents a damping parameter. Moreover, for both of these parameters, a clear trend is observed in the sense that a decrease in thickness is accompanied by an increase in the damping parameter, in agreement to the predictions of the mean free path theory. Therefore, the further discussion will mainly focus on the Drude damping parameter and its thickness dependence. Note further that in the case of gold, plasma frequencies obtained from our fits appear to be somewhat higher than the theoretical value from a previous study [15], but corresponding data reported in other studies [25] also tend to exceed that value (see also Table 2 in this regard). In that connection, it is worth noting the large scatter in reported effective mass values of conduction electrons in gold [25]. For copper, the ν_p - data obtained from our study are well below the theoretical value, in agreement with the lower density of the films.

5. Discussion

In Figure 4, the obtained model parameters $\lambda_{a/b} = 1/\nu_{a/b}$ are visualized in relation to characteristic features in the gold and copper bulk reflection spectra (Figure 4a) and in the imaginary part of the dielectric function (Figure 4b) [24]. Obviously, the parameters λ_a mark the onset wavelength of strong interband absorption structures in the spectra, while $\lambda_b - \lambda_a$ gives some effective width (in wavelength units) of the modeled interband absorption structure. Note that the ν_a values as obtained from our fits (Tables 3 and 4) are consistent with the theoretical ν_{a0} data (Table 1) in the sense that $\nu_{a0} \approx \nu_a - \Gamma$ is fulfilled.

Based on Equation (11) the thickness dependence of the obtained damping parameters is fitted by the simple law:

$$\Gamma_D(d) = \Gamma_{D,b}\left[1 + \frac{\delta}{d}\right] \qquad (13)$$

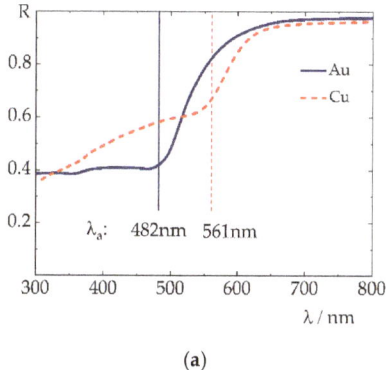
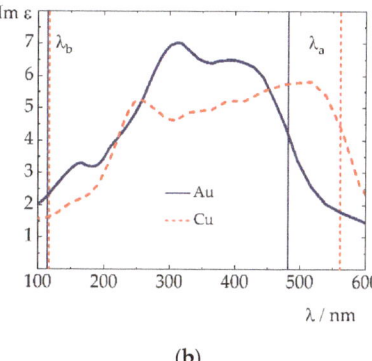

(a)   (b)

Figure 4. (a): Bulk reflectance of copper and gold. The corresponding λ_a values are indicated by vertical lines; (b): Imaginary parts of the dielectric function of gold and copper. The corresponding $\lambda_{a/b}$ values are indicated by vertical lines.

Here $\Gamma_{D,b}$ and δ are fitting parameters and the obtained values are given in Table 6. Figure 5 shows the obtained thickness dependence of Drude damping parameters together with the fit in terms of (Equation (13)). Note that our Γ_D data are principally consistent with the experimental data given in Table 2, but show a clear dependence on the film thickness.

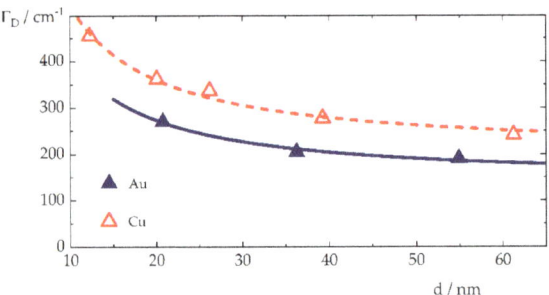

Figure 5. Thickness dependence of the Drude damping parameter Γ_D. Symbols represent data from Tables 3 and 4, while lines represent the fit in terms of Equation (13).

When now combining Equations (4), (5), (11), and (13), the following data may be calculated from the fit of $\Gamma_D(d)$:

$$\tau_b = \tau_{b0}\frac{\Gamma_{D,b0}}{\Gamma_{D,b}}; \quad (1-p_{spec}) = \frac{\delta}{2v_F\tau_b} \qquad (14)$$

The results are presented in Table 6. Also included are relaxation times τ_{J-C} obtained in previous studies [19,26].

Table 6. Parameters in the mean free path theory (Equation (11)).

Metal	τ_{b0}/fs	$\Gamma_{D,b0}$/cm^{-1}	$\Gamma_{D,b}$/cm^{-1}	τ_b/fs	τ_{J-C}/fs [26]	$v_F\tau_{b0}$/nm	$v_F\tau_b$/nm	δ/nm	$(1-p_{spec})$	p_{spec}
Cu	36	74	197.4	13.5	6.9	39.9	15.0	16.5	0.55	0.45
Au	27.3	97	135.5	19.5	9.3	37.7	26.9	20.3	0.38	0.62

Our experimental data are, thus, consistent with Equation (11) when assuming that in the Cu films, a larger amount of electrons (55%) is diffusely scattered at the film surface than in the case of gold films (38%). This result is at least in a qualitative agreement with the SEM characterization result (Figure 1), where the Cu films show a rather granular surface morphology, while in the gold films, large slabs have been found that appear rather smooth in the SEM image. Also, in the gold films, obtained damping parameters and plasma frequencies are generally closer to the theoretical values of the perfect crystal (Table 1) than in the case of copper films.

In a more refined treatment, the parameter p_{spec} could also be treated as a function of the thickness, in order to account for the larger slabs in thicker gold films, as obtained from the SEM images, but in our study the number of available experimental points is by far too small for justifying such a model extension.

Obviously, the extrapolation of our data to large thicknesses does not result in damping parameters that converge to the "perfect" values summarized in Table 1. Therefore, we have to recognize that in our films, there are at least two effects that make the Drude damping parameter different from the "perfect" values presented in Table 1:

- A thickness dependence that may obviously be described by means of the free path theory (Equation (11))
- A complicated microstructure of the film materials that results in $\tau_b < \tau_{b0}$. This is also clearly evident from the data τ_{J-C} from a previous study [26]

The complicated film microstructure as it may be guessed from the SEM images raises the question of how far simulations of charge transport in three-dimensional (3D) random microstructures (compare with Stenzel et al. [27]) may provide a theoretical access to the bulk collision times $\tau_b < \tau_{b0}$ we expect from our treatment. Although such a simulation is beyond the scope of the present study, it makes sense to compare our results, at least qualitatively, to what is obtained from such simulations. Indeed, the classical (Drude) expression for the static electric conductivity of a continuous medium may be written as:

$$\sigma_0 = \frac{N_0 e^2 \tau_{b0}}{m^*} \quad (15)$$

Here N_0 is the free electron concentration in the perfect crystal, e the elementary charge, and m^* the optical mass (compare with Ehrenreich and Philipp [15]). On the other hand, from the numerical analysis of random 3D microstructures, it turns out that their electric conductivity is determined by three crucial factors characterizing the geometry of the random network, namely:

- The filling factor $p \leq 1$;
- The constrictivity $c_r \leq 1$;
- The tortuosity $t_r \geq 1$.

Then, according to an earlier study [27], the effective conductivity of the 3D microstructure can be expressed in equations of the type:

$$\sigma_{eff} \approx \sigma_0 p^{1.0} \frac{c_r^{0.36}}{t_r^{5.17}} \quad (16)$$

with exponents that originate from a fit and vary from source to source. From $\sigma_0 = \frac{N_0 e^2 \tau_{b0}}{m^*}$ we then have:

$$\sigma_{eff} \approx \sigma_0 p \frac{c_r^{0.36}}{t_r^{5.17}} = \frac{N_0 e^2 \tau_{b0}}{m^*} p \frac{c_r^{0.36}}{t_r^{5.17}} = \frac{N_{eff} e^2 \tau_{eff}}{m^*} \quad (17)$$

When assuming identical optical masses at both sides of the above equation, the latter allows the empirical introduction of effective carrier concentrations N_{eff} and collision times τ_{eff} according to:

$$N_{eff} \approx N_0 p \leq N_0; \quad \tau_{eff} \approx \frac{c_r^{0.36}}{t_r^{5.17}} \tau_{b0} \leq \tau_{b0} \text{ because } t_r \geq 1 \text{ and } c_r \leq 1 \quad (18)$$

When finally associating τ_{eff} with the bulk relaxation time in the non-perfect film τ_b, we have

$$\tau_{eff} \approx \tau_b \leq \tau_{b0} \Rightarrow \Gamma_{D,b} = \frac{1}{4\pi c \tau_b} \geq \Gamma_{D,b0} \quad (19)$$

Therefore, the bulk Drude damping parameter in a real film shall exceed that in the perfect crystal, in agreement with our findings.

6. Summary and Outlook

We have demonstrated that the optical spectra of rather thin metal films may be reliably fitted in terms of a dispersion approach that combines the classical Drude function with the β_do model. The number of fitting parameters is given by seven dispersion model parameters plus the film thickness (es). This has allowed fitting of the metal film spectra in the wavelength range from 350 to 2000 nm. Nevertheless, inclusion of additional measurement data (e.g., ellipsometry and photometry at different angles) is known to be useful for characterization of thin metal films [28], but was not available for this study.

The fits result in a clear thickness dependence of the Drude damping parameters, which is consistent with the predictions of the mean free path theory.

In addition to the obtained thickness dependence of the Drude damping parameters, our results indicate a larger bulk value of the damping parameter than in the perfect crystal. This is consistent with earlier reports and clearly a result of the complicated microstructure of real evaporated metal films. In that connection, recent reports on the simulation of charge transport in random microstructures indicates that such simulations might give access to a numerical modeling of the complex of electrical and optical properties of real metal films of different thickness.

When returning to the application questions raised in the introduction of this paper, we conclude that a systematic increase in the Drude damping parameters with decreasing film thickness is clearly observed in our coatings and may be modeled by means of the mean free path theory, at least in the thickness range between 10 and 50 nm. However, as is evident from Figure 3, the thickness effect on the optical constants is nevertheless smaller than the range of bulk metal optical constants reported in a previous study [24], at least in the near infrared. As such, the limited reproducibility of the optical properties of copper and gold samples and the limited relevance of applied characterization methods and underlying dispersion models seems to be another important source of uncertainty in defining suitable optical constants of thin metal films in optical coating practice. In our opinion, the application of robust Kramers-Kronig consistent dispersion models with a minimum number of free parameters is important for reliable modeling of the optical constants of thin metal films. Combined with the mean free path approach, this could be an important step for modeling thickness-dependent optical constants of ultrathin metal films in optical coating practice.

Author Contributions: Conceptualization, O.S.; Data Curation, S.W.; Formal Analysis, O.S. and S.W.; Investigation, S.S., D.G. and S.-J.W.; Methodology, O.S. and S.W.; Project Administration, O.S.; Software, S.W.; Visualization, O.S., S.W., S.S., D.G. and S.-J.W.; Writing—Original Draft Preparation, O.S.; Writing—Review and Editing, S.W.

Funding: The study was partially funded by Fraunhofer internal projects.

Acknowledgments: The authors would like to thanks the attendees of the SPIE conference "Advances in Optical Coatings VI" (SPIE conf. 10691) for the ad hoc discussion on current problems in the theoretical description of ultrathin metal films optical properties, which was motivating for the present study.

Conflicts of Interest: The authors declare no conflict of interests.

References

1. Born, M.; Wolf, E. *Principles of Optics*; Pergamon Press: Oxford, UK, 1968.
2. Stenzel, O. *The Physics of Thin Film Optical Spectra: An Introduction*, 2nd ed.; Springer: Berlin, Germany, 2016; pp. 25–39.
3. Jellison, G.E. Spectroscopic ellipsometry data analysis: Measured versus calculated quantities. *Thin Solid Films* **1998**, *313*, 33–39. [CrossRef]
4. Ferlauto, A.S.; Ferreira, G.M.; Pearce, J.M.; Wronski, C.R.; Collins, R.W.; Deng, X.; Ganguly, G. Analytical model for the optical functions of amorphous semiconductors from the near-infrared to ultraviolet: Applications in thin film photovoltaics. *J. Appl. Phys.* **2002**, *92*, 2424–2436. [CrossRef]
5. Brendel, R.; Bormann, D. An infrared dielectric function model for amorphous solids. *J. Appl. Phys.* **1992**, *71*, 1–6. [CrossRef]
6. Orosco, J.; Coimbra, C.F.M. On a causal dispersion model for the optical properties of metals. *Appl. Opt.* **2018**, *57*, 5333–5347. [CrossRef]
7. Wilbrandt, S.; Stenzel, O. Empirical extension to the multioscillator model: The beta-distributed oscillator model. *Appl. Opt.* **2017**, *56*, 9892–9899. [CrossRef]
8. Macleod, H.A. *Thin-Film Optical Filters*, 4th ed.; CRC Press: Boca Raton, FL, USA, 2010.
9. Gläser, H.J. *Dünnfilmtechnologie auf Flachglas*; Verlag Karl Hofmann: Schorndorf, Germany, 1999; pp. 164–248.
10. Arndt, D.P.; Azzam, R.M.A.; Bennett, J.M.; Borgogno, J.P.; Carniglia, C.K.; Case, W.E.; Dobrowolski, J.A.; Gibson, U.J.; Hart, T.T.; Ho, F.C.; et al. Multiple determination of the optical constants of thin-film coating materials. *Appl. Opt.* **1984**, *23*, 3571–3596. [CrossRef] [PubMed]
11. Anderson, J.C. Conduction in thin semiconductor films. *Adv. Phys.* **1970**, *19*, 311–338. [CrossRef]
12. Weißmantel, C.; Hamann, C. *Grundlagen der Festkörperphysik*; Springer: Berlin, Germany, 1979; pp. 413–416.
13. Kreibig, U. Optics of Nanosized metals. In *Handbook of Optical Properties II: Optics of Small Particles, Interfaces, and Surfaces*; Hummel, R.E., Wißmann, P., Eds.; CRC Press Inc.: Boca Raton, FL, USA, 1997; pp. 145–190.
14. Gall, D. Electron mean free path in elemental metals. *J. Appl. Phys.* **2016**, *119*, 085101. [CrossRef]
15. Cooper, B.R.; Ehrenreich, H.; Philipp, H.R. Optical properties of noble metals II. *Phys. Rev.* **1965**, *138*, A494–A507. [CrossRef]
16. Ehrenreich, H.; Philipp, H.R. Optical properties of Ag and Cu. *Phys. Rev.* **1962**, *128*, 1622–1629. [CrossRef]

17. Ordal, M.A.; Bell, R.J.; Alexander, R.W.; Long, L.L.; Querry, M.R. Optical properties of fourteen metals in the infrared and far infrared: Al, Co, Cu, Au, Fe, Pb, Mo, Ni, Pd, Pt, Ag, Ti, V., and W. *Appl. Opt.* **1985**, *24*, 4493–4499. [CrossRef] [PubMed]
18. Zeman, E.J.; Schatz, G.C. An accurate electromagnetic theory study of surface enhancement factors for silver, gold, copper, lithium, sodium, aluminum, gallium, indium, zinc, and cadmium. *J. Phys. Chem.* **1987**, *91*, 634–643. [CrossRef]
19. Grady, N.K.; Halas, N.J.; Nordlander, P. Influence of dielectric function properties on the optical response of plasmon resonant metallic nanoparticles. *Chem. Phys. Lett.* **2004**, *399*, 167–171. [CrossRef]
20. Berciaud, S.; Cognet, L.; Tamarat, P.; Lounis, B. Observation of intrinsic size effects in the optical response of individual gold nanoparticles. *Nano Lett.* **2005**, *5*, 515–518. [CrossRef] [PubMed]
21. Blaber, M.G.; Arnold, M.D.; Ford, M.J. Search for the ideal plasmonic nanoshell: The effects of surface scattering and alternatives to gold and silver. *J. Phys. Chem. C* **2009**, *113*, 3041–3045. [CrossRef]
22. Stenzel, O.; Macleod, A. Metal-dielectric composite optical coatings: Underlying physics, main models, characterization, design and application aspects. *Adv. Opt. Technol.* **2012**, *1*, 463–481. [CrossRef]
23. Stenzel, O. *Optical Coatings: Material Aspects in Theory and Practice*; Springer: Berlin, Germany, 2014; pp. 307–311.
24. Palik, E.D. (Ed.) *Handbook of Optical Constants of Solids*; Academic Press: Orlando, FL, USA, 1998.
25. Olmon, R.L.; Slovick, B.; Slovick, T.W.; Shelton, S.D.; Oh, S.-H.; Boreman, G.D.; Raschke, M.B. Optical dielectric function of gold. *Phys. Rev. B* **2012**, *86*, 235147. [CrossRef]
26. Johnson, P.B.; Christy, R.W. Optical constants of the noble metals. *Phys. Rev. B* **1972**, *6*, 4370–4379. [CrossRef]
27. Stenzel, O.; Pecho, O.; Holzer, L.; Neumann, M.; Schmidt, V. Predicting effective conductivities based on geometric microstructure characteristics. *AIChE J.* **2016**, *62*, 1834–1843. [CrossRef]
28. Amotchkina, T.V.; Janicki, V.; Sancho-Parramon, J.; Tikhonravov, A.V.; Trubetskov, M.K.; Zorc, H. General approach to reliable characterization of thin metal films. *Appl. Opt.* **2011**, *50*, 1453–1464. [CrossRef] [PubMed]

© 2019 by the authors. Licensee MDPI, Basel, Switzerland. This article is an open access article distributed under the terms and conditions of the Creative Commons Attribution (CC BY) license (http://creativecommons.org/licenses/by/4.0/).

Article

Absolute Absorption Measurements in Optical Coatings by Laser Induced Deflection

Simon Bublitz and Christian Mühlig *

Department of Microscopy, Leibniz Institute of Photonic Technology, Albert-Einstein-Str. 9, 07745 Jena, Germany
* Correspondence: christian.muehlig@leibniz-ipht.de

Received: 10 July 2019; Accepted: 25 July 2019; Published: 27 July 2019

Abstract: Absolute measurement of residual absorption in optical coatings is steadily becoming more important in thin film characterization, in particular with respect to high power laser applications. A summary is given on the current ability of the laser induced deflection (LID) technique to serve sensitive photo-thermal absorption measurements combined with reliable absolute calibration based on an electrical heater approach. To account for different measurement requirements, several concepts have been derived to accordingly adapt the original LID concept. Experimental results are presented for prominent UV and deep UV laser wavelengths, covering a variety of factors that critically can influence the absorption properties in optical coatings e.g., deposition process, defects and impurities, intense laser irradiation and surface/interface engineering. The experimental findings demonstrate that by combining high sensitivity with absolute calibration, photo-thermal absorption measurements are able to be a valuable supplement for the characterization of optical thin films and coatings.

Keywords: absorption; thin films; photo-thermal technique

1. Introduction

Steadily raising power in laser material processing, finer and finer structures in semiconductor lithography and state-of-the-art optical components are faced with increasing demands and requirements. Absorption, being one of the key parameters in high-end laser applications, is gaining more and more attention due to the undesired effects resulting from thermal lensing like focus shifting, wave front deformation and depolarization. In order to take these critical issues into account in modern optic design, highly sensitive absorption measurement techniques in combination with reliable absolute calibration are required.

Commonly, measuring spectral reflectance and transmittances is the method of choice to investigate optical losses of thin films, followed by a calculation of the optical constants e.g., the refractive index n and extinction coefficient k. However, extinction coefficients obtained by that procedure contain both scatter and absorption contributions because it is not feasible to separate these parts by simply measuring transmission and reflection spectra. For particular applications, however, individual absorption and scattering data are strongly required due to their different potential interferences on the optical system performance. Further, bulk and coating/surface effects need to be separated to discriminate between different origins of absorption and scattering, respectively. Consequently, direct absorption measuring in optical thin films and bulk materials have gained more and more attention in optics qualification to ensure or improve stabile production processes, to prepare particular optical functionalities and to identify potential show stoppers upon use in high-end laser applications. Following this increasing demand, a variety of direct absorption measurement techniques have been developed recently featuring—despite particular pros and cons—a high sensitivity [1–5]. Providing not only relative but absolute absorption data has recently become more and more important e.g., in order to quantitatively simulate complex optical systems. However, with respect to an absolute calibration, the various direct absorption measurement techniques show remarkable differences when it comes to a universal and efficient procedure.

This paper will give an introduction of the laser induced deflection (LID) technique and its independent absolute calibration for the characterization of optical thin films. Further, particular LID measurement concepts are presented together with experimental results.

2. Materials and Methods

2.1. Laser Induced Deflection (LID) Technique

The laser induced deflection (LID) technique belongs to an ensemble of photo-thermal techniques with a pump-probe-configuration [6]. When the pump laser hits the sample under investigation, the absorbed pump laser power forms a temperature profile (Figure 1). The latter is turned into a refractive index profile (= thermal lens) by both the thermal expansion and the temperature dependent refractive index, respectively. The refractive index gradient accounts for a deflection of the probe beams (from the same laser source), that is proportional to the absorbed pump laser power.

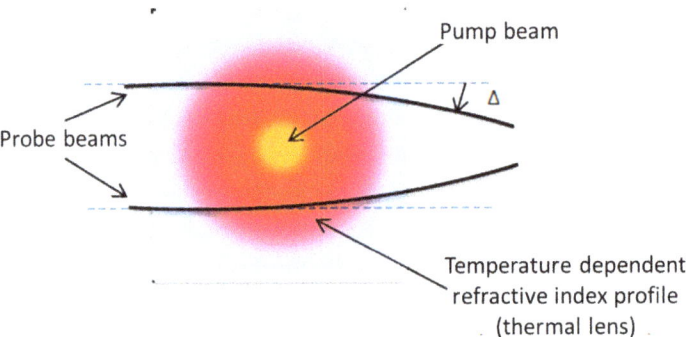

Figure 1. Sketch of the laser induced deflection (LID) measuring concept.

In the case of the LID technique, the deflection for each probe beam is detected separately by two-dimensional position sensitive detectors (PSD). The total measuring signal comprises of the sum of both PSD signals. Since the deflection directions on the PSDs have opposite signs for the two probe beams (Figure 1), the two signals are mathematically subtracted from each other. As a result, any measurement perturbations resulting from the probe laser source are strongly reduced for signal-to-noise enhancement. In combination with high averaging and the subsequent electronics, probe beam deflections in the range of some tens of Nanometers are detectable. Contrary to the majority of applied photo-thermal techniques, the LID technique uses a transversal configuration between pump and probe beam, i.e., the probe beam passes the sample under 90° to the pump beam. Thereby, in most cases crossing pump and probe beams is avoided. Instead the probe beam passes the sample outside the pump beam area. Since the refractive index profile outside the pump beam area is a function of the absorbed pump laser power only, the LID measuring signal is independent on the geometry of the pump laser beam. As another different feature compared to other photo-thermal techniques, the pump beam is not focused into the sample and typically has a larger beam size than the probe beam.

The LID measurement data is not obtained on a laser pulse-by-pulse basis. Instead, the probe beam position change with irradiation time is steadily registered with high sampling rate (16 kHz range) until it reaches its steady-state, i.e., the absorption induced refractive index gradient stays constant. To obtain a high accuracy, averaging is applied resulting in one data point every half a second. Due to the rather long data acquisition (0.5 s) and total measurement time (tens to hundreds of seconds), the deflection signal is only a function of the average absorbed pump laser power. Therefore, the actual operation mode (pulsed or continuous wave) of the pump laser does not affect the outcome of the measurement as long as the sample absorption and mean pump laser power is identical. In addition, lamps instead of laser sources can be applied as pump light source as long as the light can be shaped to an mm-size spot on the sample. Compared to alternative photo-thermal techniques, it is not required to temporally shape a pump laser operating in continuous wave mode. Furthermore, the pump laser pulse duration is not of importance for the LID measurement itself. However, the pump laser operation mode can strongly influence the sample absorption e.g., due to nonlinear absorption. This of course needs to be taken into account when choosing the pump laser source.

The LID technique itself is not limited by the pump laser wavelength as long as the average power of the pumping light is sufficient to detect the sample's absorption. A limiting factor is the wavelength of the probe laser if the investigated sample/substrate is not transparent. Here, a change of the probe beam wavelength or the use of a particular LID measuring concept (see Section 2.3.3) is required.

2.2. Absolute Calibration

Absolute calibration is a key figure of merit in photo-thermal absorption measurement techniques. For the LID technique, the approach of the calorimetry—electrical calibration—has been transferred for the first time to photo-thermal techniques. To obtain the absolute absorption value out of the LID measurement data, the thermal lens is generated by particular electric heaters. Since the shape of the thermal lens is different for bulk and coating (surface) absorption, both need to be calibrated separately for each combination of sample material and geometry [7]. In case of coating/surface absorption, small surface-mount-device (SMD) elements—fixed onto a very thin copper plate (thickness ~200 μm)—are placed centrally onto the sample surface (Figure 2). The copper plate allows for the required high thermal conduction to the sample. This calibration procedure can only be applied to transversal pump-probe configurations where the probe beam is not guided through the irradiated/heated sample part.

Figure 2. Samples of different materials for surface/coating absorption calibration.

The calibration procedure itself is composed of measuring the probe beam deflection as a function of the electric power (Figure 3a). Plotting the deflection signals versus electrical power gives a linear function (Figure 3b) which spans over several orders of magnitude for the electric power [8].

The calibration coefficient F_{CAL} is defined by the slope of the linear function including the zero-point, i.e., no electrical power means no probe beam deflection. It is valid for the applied combination of sample material and geometry. The unique feature and key advantage of this electric heating approach is the ability to calibrate the measurement setup without knowledge of the thermo-optical parameters of the sample. To verify the calibration procedure, separate reflectance, transmission, absorption and scattering measurements have been performed for different materials and coatings to proof the energy balance. The results have confirmed that within measurement accuracy, a value of 1 has been obtained for the energy balance in each of the investigations [9].

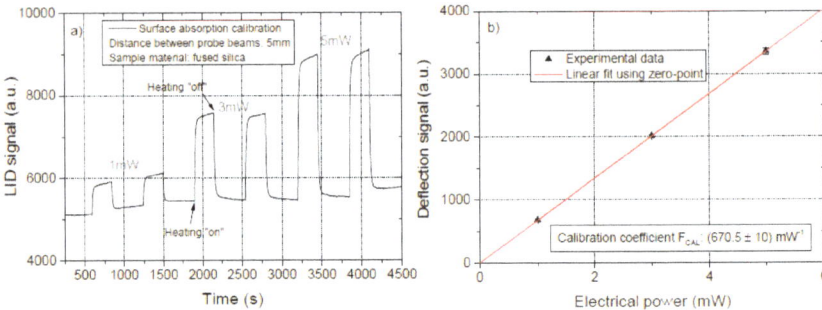

Figure 3. (a) Signal from position sensitive detector (PSD) during electrical calibration procedure for surface absorption measurement on fused silica samples of geometry ⌀ = 1″, thickness 1 mm. (b) Deflection signal as function of the electrical power for the calibration procedure in Figure 3a. The slope of the linear fitting defines the calibration coefficient F_{CAL}.

After completing the electrical calibration, the sample(s) of investigation are characterized. Figure 4 gives a measurement example for an HfO_2 single layer upon 266 nm laser irradiation.

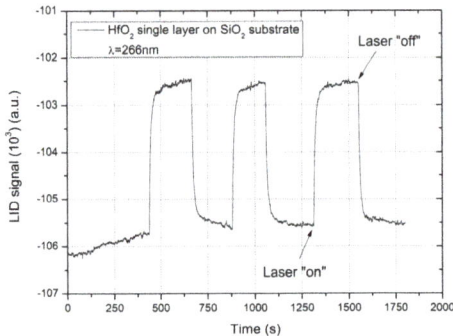

Figure 4. Experimental results for the measurement of a single HfO_2 layer onto a fused silica substrate at the laser wavelength of 266 nm.

The measurement consists of three individual cycles comprising of the time between laser irradiation "on" and "off" events. Before the first cycle of the sample irradiation starts, the "base line" of the LID signal needs to be constant or show a constant drift (Figure 4). This drift behavior is due to environmental changes like temperature in the setup surroundings. Small temperature variations already affect optic mountings within the setup by thermal expansion. As a result, small changes in the probe beam path and thus its location on the position sensitive detectors are detected.

Constant drifts during a measurement cycle are taken into account during data analysis. Here, the LID deflection signal is defined as the difference between the "base line" signal and the steady-state signal for the constant thermal lens upon irradiation. Once the LID deflection signals for all measurement cycles are obtained, their mean value is defined as the LID intensity I_{LID}. Together with the corresponding mean pump laser power P_L for all cycles and the calibration coefficient F_{CAL}, the coating absorption A is calculated by

$$A = I_{LID}/(F_{CAL} \times P_L) \tag{1}$$

2.3. LID Measurement Concepts

In order to account for the variety of measurement tasks, different concepts based on the LID technique have been developed (Figure 5). For the different concepts, it was investigated whether to pass two probe beams through the sample outside the irradiated area or just one probe beam inside the irradiated area. Further, the two-dimensional PSD allows detecting both, the probe beam deflection in direction of the pump laser (= horizontal) or perpendicular to it (= vertical). Due to the general scheme in Figure 1 it was tested how the LID technique can be adapted to round sample geometries without side-face polishing, which is a preferred geometry in coating manufacturing. Finally, particular attention is paid to increasing the sensitivity of substrates that are disadvantageous for photo-thermal measurements (e.g., Sapphire, CaF_2 ...). However, all the different concepts have been considered with respect to maintaining the electrical calibration procedure.

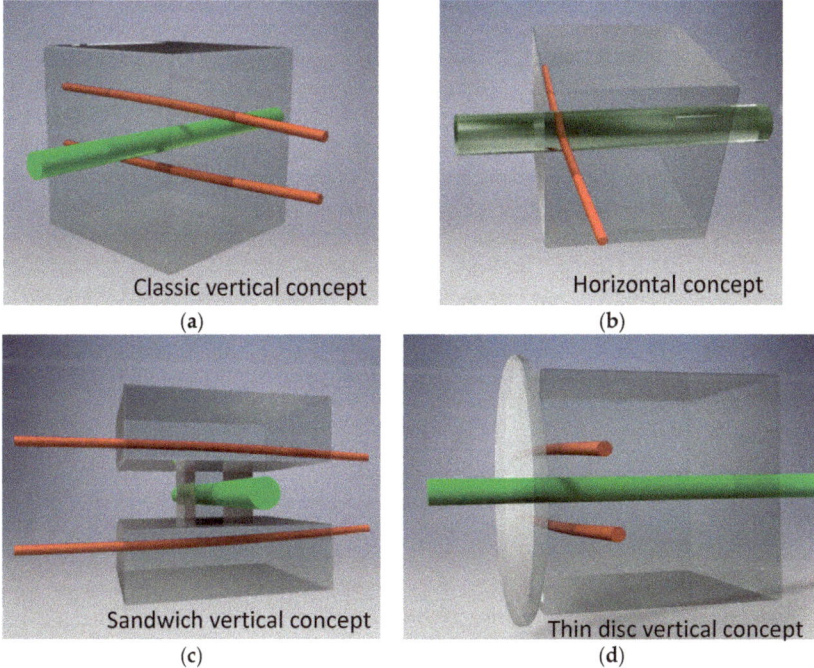

Figure 5. Different measurement concepts based on the LID technique: (**a**) Classic vertical concept, (**b**) horizontal concept, (**c**) Sandwich vertical concept and (**d**) Thin disc vertical concept. In all schemes, the green beam assigns the pump beam whereas the red beam(s) represent the probe beam(s).

2.3.1. Classic Vertical Concept

The classic vertical concept (Figure 5a) applies two probe beams above/beneath the irradiated spot and utilizes the probe beam deflection perpendicular to the pump beam direction. A high signal-to-noise ratio makes the vertical concept the best option for absorption measurements in bulk and highly reflective coatings if the required sample geometry can be supplied. The position of the probe beam can be changed with respect to the sample length. Consequently, for measuring coatings the probe beams pass the sample close to the coated surface while for bulk absorption measurement the probe beams pass the sample in the middle of the sample length where influences of the front/rear surfaces are minimal. In the case of transparent coatings, the deflection always comprises of both, the coating and substrate absorption. For separation, there are two approaches. For the first, an uncoated reference substrate of same geometry/material is measured additionally, and the difference in the deflection signals is assigned to the coating absorption. In case of double-side coated surfaces, however, only the sum of both coating absorptions is accessible. In order to avoid systematic errors, it is essential that the coated and uncoated substrates feature identical bulk and surface absorption properties. For a second approach, the deflection signals are measured at different positions along the sample length. In case of a single-side coated sample (neglecting the absorption of the uncoated surface), two different positions are required. Together with the corresponding bulk and surface absorption calibrations, the bulk and coating absorptions can be calculated. In contrast to the first approach, the individual coating absorptions of a double-side coated substrate are detectable by using three different measurement positions along the sample length.

2.3.2. Horizontal Concept

The only concept that breaks with the general LID scheme is the horizontal concept for coating measurements only. Here, in contrast, only one probe beam is guided centrally through the middle of the irradiated area at a position closest to the coated sample surface (Figure 5b). The resulting deflection in the direction of the pump beam allows for detecting the surface absorption virtually free of the sample's bulk absorption [9]. Therefore, the horizontal concept is an alternative choice in particular for investigating anti-reflecting or partially reflecting coatings, where typically both bulk and coating absorption are present in the measurement signal for the classic vertical concept. Using only one probe beam instead of two reduces the signal-to-noise ratio compared to the classic vertical concept. However, no separation between bulk and coating absorption is required which yields an overall improvement for the measurement accuracy. A limitation for absolute absorption measurements is given by guiding the probe beam through the pump laser volume. The use of the electrical calibration now requires virtually identical areas for electrical and laser heating. This limits the concept to top-hat shaped pump laser beams.

2.3.3. Sandwich Vertical Concept

The sensitivity of photo-thermal absorption measurement techniques depends on the thermo-optical properties of the investigated material. For a variety of interesting materials, e.g., crystals like Sapphire or CaF_2, this results in very small thermal lenses. While this is favorable in the application, the measurement sensitivity easily gets insufficient for very low absorption detection.

The recently demonstrated sandwich vertical concept (Figure 5c) solves this issue [10]. The basic principle consists of a small sample of investigation that is placed between two larger tiles of an appropriate optical material ($\rightarrow$ sandwich). The pump laser still hits the sample of investigation and generates the absorption induced heat. The probe beams, however, are passing the sandwich tiles instead of the sample. The probe beams are now deflected by the thermal lens that is generated in the optical tiles after the absorption induced heat in the sample is transferred into the tiles.

There are a few prominent features related to this concept. Compared to the LID vertical concept, it enables a strong increase in measurement sensitivity for materials with a low photo-thermal response by choosing appropriate optical tile materials, e.g., fused silica [10]. The reason is that now the probe beams use the thermo-optical properties of the sandwich tile material instead of those of the sample material. Strictly speaking, the low sensitivity materials like Sapphire or CaF_2 are now measured with the sensitivity for fused silica. Separating the sites of pump beam absorption (sample of investigation) and probe beam deflection (optical tiles) allows a sensitivity increase for many optical materials by more than an order of magnitude.

Apart from the sensitivity increase, the Sandwich vertical concept also overcomes another drawback of the LID technique. As a result of the transversal pump-probe configuration, not only two, but in total four polished side-faces are required in order to allow both pump and probe beam guiding through the sample. Recently, it was proven by electrical calibration and real measurements that the amount of heat transferred into the optical tiles until stationary thermal lens condition is the same whether the contact surfaces are polished or not [11]. Therefore, the requirement for additional side-face polishing is omitted. Moreover, the investigated material does not need to be transparent for the probe beam anymore. This clears the way for potential investigations of infrared optical materials or metallic substrates.

2.3.4. Thin Disc Vertical Concept

Commonly in optical coating manufacturing, round shaped substrates are standard for thin film investigations. In particular, substrates of 1" diameter and 1mm thickness are used as reference samples for post-coating characterizations. In order to match this geometry to the LID concept, the thin disc vertical concept has been developed (Figure 5d). Using a particularly designed holder, a thin round sample (1–2" diameter, thickness: 1–2 mm) is pressed against a rectangular (cubic) substrate. The latter possesses at least four polished side-faces to allow for a measurement similar to the classic vertical concept. Similar to the Sandwich vertical concept, the probe beams are guided through the rectangular substrate—instead of using the actual sample of investigation—and use the thermal lens that is generated by heat conduction from the absorption in the coated thin sample. For transparent coating investigations, the LID measuring signal is composed of three absorptions originating from the coating itself, the thin round sample substrate and the rectangular substrate. To identify the coating absorption, it is required to perform a reference measurement using an uncoated thin substrate of same geometry/material attached to the rectangular substrate. In case of double-side coated surfaces, however, only the sum of both coating absorptions is accessible. It is essential that the thin substrates feature identical bulk and surface absorption properties in order to avoid systematic errors.

3. Results and Discussion

This section contains a variety of experimental results in order to show the benefit of using direct absorption measurements for optical coating characterization. The scope ranges from influences of deposition parameters or impurities to laser irradiation induced effect, nonlinear absorption and separation between thin film and interface absorption.

3.1. Absorption vs. Coating Deposition Temperature (Thin Disc Vertical Concept)

One of the critical parameters in optical coating manufacturing is the deposition temperature. It has, for example, an impact on the thin film structure ($\rightarrow$ refractive index, scattering) or the stoichiometry ($\rightarrow$ defects). Nonstoichiometric films often give rise to increased absorption due to the formation of intrinsic defects. For oxide coatings these are preferable oxygen vacancies [12–14] whereas for fluorides the formation of color centers or even localized metallic clusters is critical [15,16]. For our studies, high-reflecting (HR) mirrors (MgF_2/LaF_3) for 193 nm and AOI = 45° have been deposited onto CaF_2 substrates (1" diameter, thickness 2 mm) by thermal evaporation using deposition temperatures in the range 250–370 °C. Absorption measurements have been carried out by means of the thin disc vertical concept using an ArF excimer laser (ExciStar S Industrial, Coherent GmbH, Goettingen, Germany) with a repetition rate f = 1 kHz. Upon laser irradiation, all samples showed an initial absorption decrease related to laser induced desorption of hydrocarbons (Figure 6a) [17]. Figure 6b displays the absorption data after initial decrease as function of the deposition temperature. The data show a linear increase in the HR coating absorption with deposition temperature.

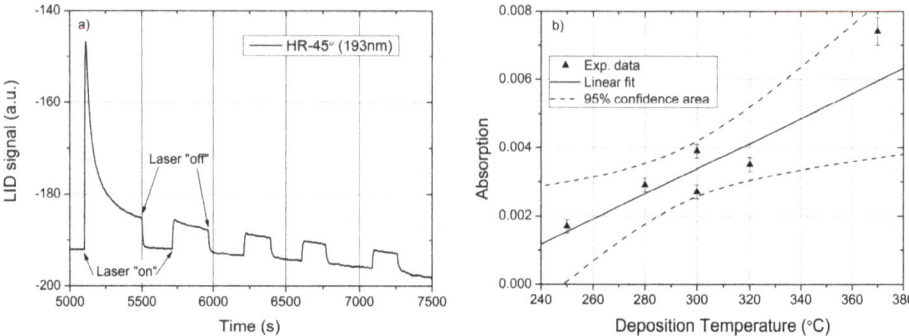

Figure 6. (a) Laser induced absorption decrease in HR coatings (MgF$_2$/LaF$_3$) upon initial ArF laser irradiation due to hydrocarbon desorption, (b) Increasing HR coating absorption with deposition temperature. Absorption data are obtained after complete hydrocarbon desorption. (Adapted from [18]. Copyright 2008 SPIE International.)

However, it cannot be distinguished whether both materials contribute to the raising absorption. The finding of an absorption increase is somewhat unexpected, since often higher temperature (annealing) yields absorption improvement in optical thin films like LaF$_3$ [19]. However, interface or structural properties (layer growth) may strongly depend on the deposition process and temperature and affect the overall absorption. Further, temperature related changes may be advantageous or disadvantageous, respectively, for the individual coating materials.

3.2. Absorption vs. Impurities (Thin Disc Vertical Concept)

Besides intrinsic defects, impurities play a large role for the absorption properties of optical coatings. However, they are caused by other sources like insufficiently pure raw materials or inner parts of the deposition chamber, e.g., crucibles. We compared different HfO$_2$ single layers (optical thickness: 4λ/4 @266 nm) on fused silica substrates (1″ diameter, thickness 1mm), made by ion-assisted deposition (IAD) from different raw materials. Since HfO$_2$ is the most preferred high-index material in the UV range, we tested the absorption at the prominent laser wavelengths 355 nm (Spruce-355/5, Huaray Precision Lasers, Wuhan, China) and 266 nm (NANIO 266-2-V-development, InnoLas Photonics GmbH, Krailling, Germany), respectively. All thin films have been characterized by TOF-SIMS analysis. It was found, that in particular the iron content strongly varies amongst the raw materials. Figure 7 shows the HfO$_2$ layer absorption as a function of the iron content. It is seen that the HfO$_2$ absorption at 355 nm is significantly lower than at 266 nm which is attributed to the cumulative influence of the HfO$_2$ band edge. For both wavelengths, the absorption increases linearly with the iron concentration which coincides with the common very broad absorption bands of metallic impurities.

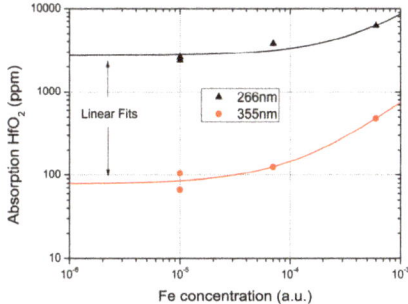

Figure 7. Absorption in HfO$_2$ single layers as a function of the iron concentration for the wavelengths 266 and 355 nm, respectively.

3.3. Absorption Change upon Initial Laser Irradiation (Thin Disc Vertical + Sandwich Concept)

Providing invariant laser operation requires optical components that do not alter upon laser irradiation. However, it is well-known that under exposure to UV or DUV laser light, even intrinsic defects in highly transparent optical materials can alter. A prominent example is fused silica where defect generation, defect annealing and defect transformation can occur upon initial ArF laser irradiation [20]. Hence it is worth looking to the absorption properties of optical coatings with respect to their changes under intense UV or DUV laser irradiation. An AR coating based on MgF$_2$/LaF$_3$ has been deposited by thermal evaporation onto a CaF$_2$ substrate (1″ diameter, thickness 1 mm). In addition, each of the coating materials has been deposited individually on separate substrates within the coating runs. The CaF$_2$ substrates have been attached to a fused silica cube for measurement (see Figure 5d). Combining the thin vertical concept with the idea of the sandwich concept allows for a strong sensitivity increase (~factor of 10) by using the photo-thermal properties of fused silica instead of CaF$_2$. The three samples (AR, LaF$_3$, and MgF$_2$) have been irradiated at 266 nm (NANIO 266-2-V-development, InnoLas Photonics GmbH) with intermediate absorption measurements. Figure 8a shows that during 10 minutes of laser irradiation, the absorption of the AR coating roughly has doubled before reaching a stationary value. The results of the individual coating materials give proof that while the LaF$_3$ thin film does not show any irradiation induced absorption change, the MgF$_2$ thin film nearly tripled its absorption during initial irradiation (Figure 8b).

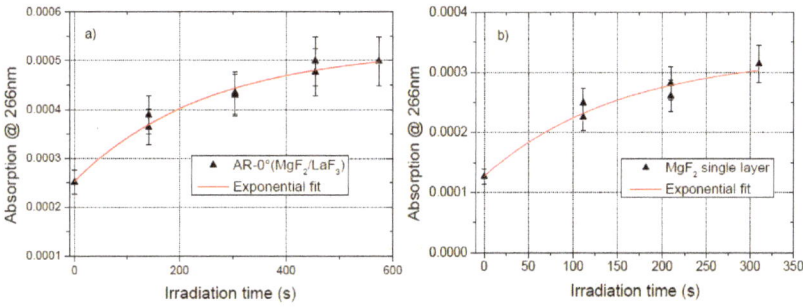

Figure 8. Absorption vs. laser irradiation time at 266 nm for (**a**) an AR-0° coating based on MgF$_2$/LaF$_3$ thin films, (**b**) an MgF$_2$ single layer deposited identically to the MgF$_2$ part of the AR coating.

3.4. Separation between Thin Film and Interface Absorption (Thin Disc Vertical + Sandwich Concept)

For transparent optical thin films like AR coatings, interface engineering i.e., surface polishing and cleaning, is the key to low absorption and corresponding to high resistance against very intense laser irradiation (LID—laser induced damage threshold). Inadequate surface preparation before the coating process may result in surface absorption not caused by the thin film itself. Consequently, strongly varying interface/surface absorption is likely to occur by different polishing/cleaning/coating technologies. Further, in order to simulate a total absorption of HR coatings, the absorption of the individual layer materials is required without contributions related to the substrate surface. Hence, separating between interface and thin film absorption by appropriate experiment design is highly required. One way to solve that issue is measuring a series of samples with different optical film thicknesses but identical surface preparation. Figure 9 gives the results for LID absorption measurements at 266 nm (NANIO 266-2-V-development, InnoLas Photonics GmbH, Krailling, Germany) of single MgF_2 and LaF_3 single layers (thermal evaporation) on CaF2 substrates (1″ diameter, thickness 1 mm) which have been attached to a fused silica cube for sensitivity enhancement.

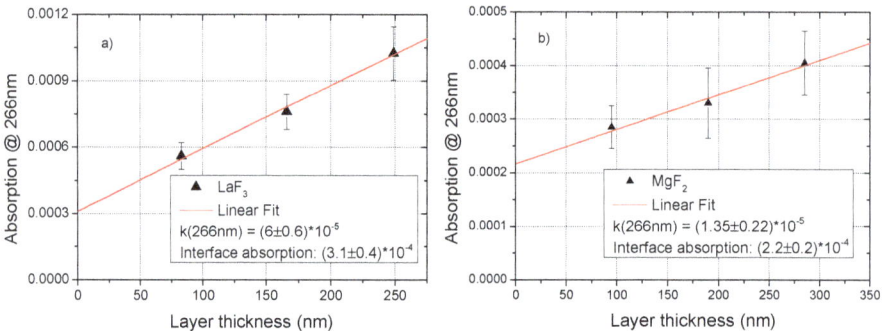

Figure 9. Absorption at 266 nm vs. thin film layer thickness and calculated interface and layer absorption k, respectively, of (**a**) LaF_3 thin films and (**b**) MgF_2 thin films.

The different layer thicknesses are matched to allow for optical thickness of multiple of $\lambda/4$ in order to generate identical surface reflectance. The results in Figure 9 prove that the interface absorption contributes significantly to the total surface absorption. In particular for the thinnest layer of $\lambda/4$ thickness, the interface contribution can reach up to 50% and higher of the overall absorption. Consequently, the simulation for an HR mirror based on the investigated MgF_2 and LaF_3 materials would considerably overestimate its absorption and underestimate the achievable reflectivity. The k values, calculated by

$$k = (\alpha \times \lambda)/4\pi \qquad (2)$$

with α being the absorption coefficient per Nanometer and λ the wavelength, are in the 10^{-5} range. These are reasonable data compared to commonly published data for the deep UV range [15,21].

3.5. Nonlinear Thin Film Absorption (Horizontal Concept)

For high laser intensities and/or short laser wavelengths, not only the linear but also a potential nonlinear absorption in the coatings need to be considered.

In the case of bulk materials, optical materials like SiO_2 or metal fluorides show very small intrinsic absorption due to their large bandgaps. High laser intensities, however, enables intrinsic or defect related nonlinear absorption, which is not accessible to common transmission measurements using spectrophotometers. Consequently, direct absorption measurement techniques combined with intense laser light have been applied to investigate the nonlinear absorption behavior in DUV bulk materials, mainly at the wavelengths 193, 248 and 266 nm. In contrast, very few experimental results exist with respect to the nonlinear absorption of the common thin film materials for the DUV wavelength region. This might originate from the linear absorption coefficients of coatings, which are significantly higher than for the corresponding bulk materials. Typically, any optical coating of a thickness of 10 mm would be completely opaque. Therefore, small nonlinear absorption might be hidden by a strong linear absorption background. However, recent experiments using direct absorption measurements have revealed measurable nonlinear absorption even in thin single layers [22–24]. Similar to the linear absorption, it was demonstrated that two-photon absorption (TPA) coefficients in optical coatings remarkably exceed those for the corresponding bulk materials. To a certain extent, these enlarged nonlinear absorption values are referred to sequential two-step absorption processes via intermediate defect energy levels [22].

For investigating nonlinear absorption properties, two different MgF_2 thin films (thermal evaporation, Table 1) on CaF_2 substrates (15 × 15 × 10 mm^3) have been measured at 193nm (ExciStar S Industrial, Coherent GmbH, Goettingen, Germany) as a function of the laser intensity. For very weak absorption values, i.e., $\alpha h \ll 1$ and $\beta I_0 h \ll 1$, the light absorption A in a thin film of a thickness h as a function of the incident laser intensity I_0 can be simplified to

$$A = (\alpha + \beta \times I_0) \times h \tag{3}$$

where α and β denote the one- and two-photon absorption coefficient, respectively. Therefore, using Equation (3) for the results of thin film absorption measurements as function of the laser intensity (or fluence) allow for a calculation of the two-photon absorption coefficient β. Figure 10 shows the results for the two MgF_2 thin films.

Table 1. Linear absorption k and two-photon absorption coefficient β of the investigated MgF_2 single layers.

Sample	Physical Thickness (nm)	k (H→0) [10^{-4}]	β [10^{-5} cm/W]
MgF_2 (1)	200	6.9 ± 0.35	5.1 ± 3.8
MgF_2 (2)	134	1.8 ± 0.06	1.8 ± 0.6

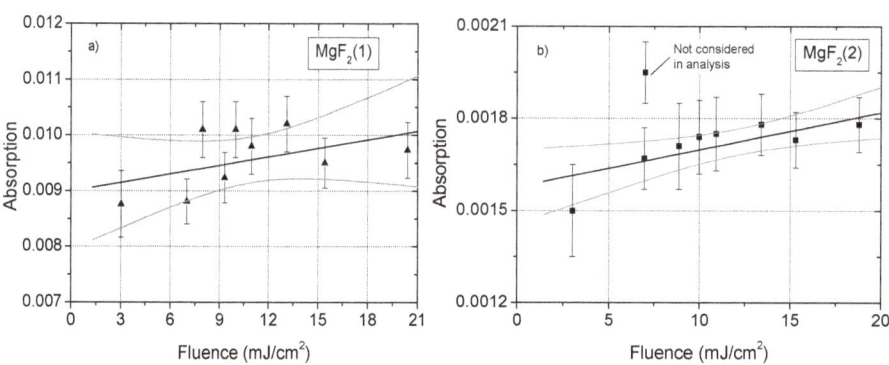

Figure 10. Absorption at 193 nm vs. laser fluence for two different MgF_2 thin films (see Table 1) (Adapted from [25]. Copyright 2009 The Optical Society).

Table 1 summarizes the calculated extinction coefficients and two-photon absorption coefficients for the investigated MgF_2 thin films. The k values in the low to medium 10–4 range are very comparable to extinction coefficients obtained recently by spectral reflectance and transmission measurements [15,16,21,26,27]. As in recently published investigations, the obtained two-photon absorption coefficients in this work are several orders of magnitude higher than those obtained for fluoride single crystals [28–31]. This might be explained in two ways. First, due to the typically low crystalline order of the thin films, their bandgaps are supposed to be lower than for the corresponding single crystal which can result in a higher nonlinear absorption coefficient. Second, a larger number of real defect states (extrinsic or intrinsic nature) are responsible for the high single photon absorption coefficient in thin films compared to the single crystals. Therefore, bridging the bandgap is likely to occur by sequential two-step absorption using intermediate defect energy levels. For such a scenario, the defect related, effective two-photon absorption can substantially exceed the intrinsic two-photon absorption via virtual energy states [31].

4. Summary and Outlook

We have demonstrated that the combination of photo-thermal absorption measurement and reliable absolute calibration via electrical heaters allow for enhanced characterization of optical coatings. Based on the general idea of the laser induced deflection (LID) technique, a variety of concepts have been derived and presented to allow for different measurement requirements. Here, particular attention has been paid to realize the measurement of round sample geometries commonly used in coating characterization. Further, measuring transparent optical coatings with negligible contribution from the underlying substrate has been demonstrated. Finally, a concept is depicted to overcome the drawback of photo-thermally insensitive substrate materials like CaF_2 or Sapphire.

Reliable absolute calibration is one of the most important issues for photo-thermal absorption measurements. In contrast to common simulation procedures or use of samples with known absorption properties, an electrical calibration adapted from the calorimetric absorption measurement is applied. Hereby, an absolute calibration of the setup is carried out without any knowledge of the substrate's material parameters. However, it is required to use a calibration substrate of same geometry and material as the sample of investigation.

Experimental results have been presented for prominent UV and deep UV laser wavelengths (355, 266 and 193 nm), covering a variety of factors that critically can influence the absorption properties in optical coatings e.g., the deposition process, defects and impurities, intense laser irradiation and surface/interface engineering. The UV and deep UV wavelength region has been chosen because here, the combination of high laser intensity and high photon energy gives rise to enhanced interaction between the laser light and the optical coating. Further, prominent absorption bands of defects and impurities are located in the considered spectral region, e.g., color centers or oxygen deficiency centers.

With respect to round substrate geometries, current activities are ongoing to adapt the LID technique also to thick round standard substrates (e.g., thickness $\frac{1}{4}$"). Applying the thin round vertical concept would result in too low a sensitivity due to the large distance between the coating and the probe beams. The thick substrates are in particular applied for mirrors of highest reflectance (>99.99%). The ability to use the LID technique would give an additional advantage to optical coating characterization in combination with the cavity ring-down (CRD) technique that is used to precisely determine the mirror's reflectance. Thereby it would be possible to separate between absorption and scattering loss contributions allowing a defined improvement of highly reflecting mirrors.

Author Contributions: Conceptualization, C.M.; Validation, S.B. and C.M.; Formal Analysis, S.B. and C.M.; Investigation, S.B and C.M; Data Curation, C.M.; Writing—Original Draft Preparation, C.M.; Writing—Review and Editing, S.B.; Visualization, C.M.; Supervision, C.M.; Project Administration, C.M.; Funding Acquisition, C.M.

Funding: This research was funded by GERMAN FEDERAL MINISTRY OF ECONOMICS AND TECHNOLOGY within the framework of the Confederation of Industrial Research Associations, grant numbers KF2206909UW2 and ZF4006807RE5.

Acknowledgments: The authors are grateful to company Layertec GmbH for providing the majority of the coated samples as well as for the TOF-SIMS investigations.

Conflicts of Interest: The authors declare no conflict of interest.

References

1. Li, B.; Blaschke, H.; Ristau, D. Combined laser calorimetry and photothermal technique for absorption measurement of optical coatings. *Appl. Opt.* **2006**, *45*, 5827–5831. [CrossRef] [PubMed]
2. Leonid, S. Light-induced absorption in materials studied by photothermal methods. *Recent Pat. Eng.* **2009**, *3*, 129–145. [CrossRef]

3. Alexandrovski, A.; Fejer, M.; Markosian, A.; Route, R. Photothermal common-path interferometry (PCI): New developments. *Proc. SPIE* **2009**, *7193*, 71930D.
4. Yoshida, S.; Reitze, D.H.; Tanner, D.B.; Mansell, J.D. Method for measuring small optical absorption coefficients with use of a Shack–Hartmann wave-front detector. *Appl. Opt.* **2003**, *42*, 4835–4840. [CrossRef] [PubMed]
5. Waasem, N.; Fieberg, S.; Hauser, J.; Gomes, G.; Haertle, D.; Kühnemann, F.; Buse, K. Photoacoustic absorption spectrometer for highly transparent dielectrics with parts-per-million sensitivity. *Rev. Sci. Instrum.* **2013**, *84*, 23109. [CrossRef] [PubMed]
6. Guntau, M.; Triebel, W. A Novel method to measure bulk absorption in optically transparent materials. *Rev. Sci. Instr.* **2000**, *71*, 2279–2282. [CrossRef]
7. Mühlig, C.; Triebel, W.; Kufert, S.; Bublitz, S. Characterization of low losses in optical thin films and materials. *Appl. Opt.* **2008**, *47*, C135–C142. [CrossRef]
8. Mühlig, C.; Kufert, S.; Bublitz, S.; Speck, U. Laser induced deflection technique for absolute thin film absorption measurement: Optimized concepts and experimental results. *Appl. Opt.* **2011**, *50*, C449–C456. [CrossRef]
9. Mühlig, C.; Bublitz, S. Sensitive and absolute absorption measurements in optical materials and coatings by laser induced deflection (LID) technique. *Opt. Eng.* **2012**, *51*, 121812. [CrossRef]
10. Mühlig, C.; Bublitz, S.; Paa, W. Enhanced laser-induced deflection measurements for low absorbing highly reflecting mirrors. *Appl. Opt.* **2014**, *53*, A16–A20. [CrossRef]
11. Mühlig, C.; Bublitz, S.; Paa, W. Sandwich concept: Enhancement for direct absorption measurements by laser induced deflection (LID) technique. In Proceedings of the Laser-Induced Damage in Optical Materials, Boulder, CO, USA, 23–26 September 2012.
12. Heber, J.; Mühlig, C.; Triebel, W.; Danz, N.; Thielsch, R.; Kaiser, N. Deep UV laser induced luminescence in oxide thin films. *Appl. Phys. A* **2002**, *75*, 637–640. [CrossRef]
13. Papernov, S.; Brunsman, M.D.; Oliver, J.B.; Hoffman, B.N.; Kozlov, A.A.; Demos, S.G.; Shvydky, A.; Cavalcante, F.H.; Yang, L.; Menoni, C.S.; et al. Optical properties of oxygen vacancies in HfO_2 thin films studied by absorption and luminescence spectroscopy. *Opt. Exp.* **2018**, *26*, 17608–17623. [CrossRef] [PubMed]
14. Ivanova, E.V.; Zamoryanskaya, M.V.; Rustovarov, V.A.; Aliev, V.S.; Gritsenko, V.A.; Yelisseyev, A.P. Cathodo-and photoluminescence increase in amorphous hafnium oxide under annealing in oxygen. *J. Exp. Theor. Phys.* **2015**, *120*, 710–715. [CrossRef]
15. Dumas, L.; Quesnel, E.; Pierre, F.; Bertin, F. Optical properties of magnesium fluoride thin films produced by argon ion-beam assisted deposition. *J. Vac. Sci. Technol. A* **2002**, *20*, 102–106. [CrossRef]
16. Larruquert, J.I.; Keski-Kuha, R.A.M. Far ultraviolet optical properties of MgF_2 films deposited by ion-beam sputtering and their application as protective coatings for Al. *Opt. Commun.* **2003**, *215*, 93–99. [CrossRef]
17. Heber, J.; Mühlig, C.; Triebel, W.; Danz, N.; Thielsch, R.; Kaiser, N. Deep UV laser induced luminescence in fluoride thin films. *Appl. Phys. A* **2003**, *76*, 123–128. [CrossRef]
18. Mühlig, C.; Triebel, W.; Kufert, S.; Bublitz, S. Laser induced fluorescence and absorption measurements for DUV optical thin film characterization. In Proceedings of the Advances in Optical Thin Films III, Scotland, UK, 2–5 September 2008.
19. Mühlig, C. Absorption and fluorescence measurements in optical coatings. In *Optical Characterization of Thin Solid Films*, 1st ed.; Stenzel, O., Ohlidal, M., Eds.; Springer International Publishing AG: Cham, Switzerland, 2018; Volume 64, pp. 407–432.
20. Mühlig, C.; Stafast, H.; Triebel, W. Generation and annealing of defects in virgin fused silica (type III) upon ArF laser irradiation: Transmission measurements and kinetic model. *J. Non-Cryst. Solids* **2008**, *354*, 25–31. [CrossRef]
21. Bischoff, M.; Gäbler, D.; Kaiser, N.; Chuvilin, A.; Kaiser, U.; Tünnermann, A. Optical and structural properties of LaF_3 thin films. *Appl. Opt.* **2008**, *47*, C157–C161. [CrossRef]
22. Apel, O.; Mann, K.; Zoellner, A.; Goetzelmann, R.; Eva, E. Nonlinear absorption of thin Al_2O_3 films at 193 nm. *Appl. Opt.* **2000**, *39*, 3165–3169. [CrossRef]
23. Li, B.; Xiong, S.; Zhang, Y.; Martin, S.; Welsch, E. Nonlinear absorption measurement of UV dielectric components by pulsed top-hat beam thermal lens. *Opt. Commun.* **2005**, *244*, 367–376. [CrossRef]
24. Jensen, L.; Mende, M.; Schrameyer, S.; Jupé, M.; Ristau, D. Role of two-photon absorption in Ta_2O_5 thin films in nanosecond laser-induced damage. *Opt. Exp.* **2012**, *37*, 4329–4331.

25. Mühlig, C.; Bublitz, S.; Kufert, S. Nonlinear absorption in single LaF$_3$ and MgF$_2$ layers at 193 nm measured by surface sensitive laser induced deflection technique. *Appl. Opt.* **2009**, *48*, 6781–6787. [CrossRef] [PubMed]
26. Taki, Y. Film structure and optical constants of magnetron-sputtered fluoridic films for deep ultraviolet lithography. *Vacuum* **2004**, *74*, 431–435. [CrossRef]
27. Shuzhen, S.; Jianda, S.; Chunyan, L.; Kui, Y.; Zhengxiu, F.; Lei, C. High-reflectance 193 nm Al$_2$O$_3$/MgF$_2$ mirrors. *Appl. Surf. Sci.* **2005**, *249*, 157–161. [CrossRef]
28. Görling, C. Wechselwirkungsmechanismen Von DUV- und VUV-Laserstrahlung Mit Fluoridischen Einkristallen. Ph.D. Thesis, Georg-August-University, Gottingen, Germany, 2003.
29. Taylor, A.J.; Gibson, R.B.; Roberts, J.P. Two-photon absorption at 248 nm in ultraviolet window materials. *Opt. Lett.* **1988**, *13*, 814–816. [CrossRef] [PubMed]
30. Kittelmann, O.; Ringling, J. Intensity-dependent transmission properties of window materials at 193 nm irradiation. *Opt. Lett.* **1994**, *19*, 2053–2055. [CrossRef] [PubMed]
31. Mühlig, C.; Triebel, W.; Stafast, H.; Letz, M. Influence of Na-related defects on ArF laser absorption in CaF$_2$. *Appl. Phys. B* **2010**, *99*, 525–533. [CrossRef]

© 2019 by the authors. Licensee MDPI, Basel, Switzerland. This article is an open access article distributed under the terms and conditions of the Creative Commons Attribution (CC BY) license (http://creativecommons.org/licenses/by/4.0/).

MDPI
St. Alban-Anlage 66
4052 Basel
Switzerland
Tel. +41 61 683 77 34
Fax +41 61 302 89 18
www.mdpi.com

MDPI Books Editorial Office
E-mail: books@mdpi.com
www.mdpi.com/books

www.ingramcontent.com/pod-product-compliance
Lightning Source LLC
LaVergne TN
LVHW071952080526
838202LV00064B/6730